William E. Cooper John M. Sorensen

Fundamental Frequency in Sentence Production

With 85 Figures

Springer-Verlag
New York Heidelberg Berlin

William E. Cooper
Department of Psychology and Social Relations
Harvard University
Cambridge, Massachusetts 02138
USA

John M. Sorensen
Bolt Beranek and Newman, Inc.
Cambridge, Massachusetts 02238
USA

Library of Congress Cataloging in Publication Data

Cooper, William E
Fundamental frequency in sentence production.

Bibliography: p.
Includes indexes.
1. Psycholinguistics. 2. Speech—Physiological aspects. 3. Psychoacoustics. I. Sorensen, John M., joint author. II. Title.
P37.C62 401′.9 80-24897

Printed in the United States of America

9 8 7 6 5 4 3 2 1

ISBN 0-387-90510-3 Springer-Verlag New York Heidelberg Berlin
ISBN 3-540-90510-3 Springer-Verlag Berlin Heidelberg New York

For Jeanne and Linda

Preface

The scientific study of speech production has been undertaken within a variety of disciplines, including linguistics, physiology, physical acoustics, and psychology, among others. In recent years, it has become possible to combine the knowledge accumulated within these separate disciplines to gain more insight into the nature of this complex human activity. Here, we apply an interdisciplinary focus to one of the most widely discussed and poorly understood aspects of the speech signal. This study of fundamental voice frequency involves a mixture of linguistics, physical acoustics, and psychology. The relevant aspects of each discipline are introduced in Chapter 1 so that readers with minimal background in at least one of these areas should be able to comprehend much of the whole.

This book is written primarily as an original research treatise rather than as a review of the literature. However, many previous studies of fundamental frequency are discussed here, and the book should serve as both a reference source and a supplementary text in courses on speech communication taught within a number of disciplines. Because of the complexity of F_0 patterns, graphical displays of the results play a particularly important role in this presentation. While the results of each experiment are summarized and discussed verbally within the text, the graphical displays, by comparison, represent a more integral part of this book than is often the case in treatments of speech production.

Within the first five chapters, new findings and conclusions are presented that seem to provide a firmer basis for understanding the production of fluent speech, particularly from the combined standpoints of cognitive psychology and physical acoustics. The empirical studies have yielded a quantitative characterization of many sentential features of fundamental frequency as well as clues about the speaker's controlling psychological computations. Although fundamental frequency is only one of many important characteristics of the speech wave, an in-depth treatment of this attribute impinges on a rich variety of issues, and a wide-ranging discussion of related topics is presented in Chapter 6.

Acoustically, fundamental frequency is very difficult to measure in a manner that is both precise and efficient, and this drawback has hampered many previous efforts. Our own research could not have been conducted at all without the help of Dr. William Henke, who wrote the computer program employed throughout virtually all the empirical studies. Dr. Henke taught JMS about the design and use of the program. In addition, he updated the program at various points during the project to suit particular needs.

Special thanks also go to D. Keith North, for performing innumerable technical duties with great care and precision; Kazuhiko Yorifuji, for his collaboration on the study of Japanese reviewed in Chapter 2; Dr. Douglas O'Shaughnessy, for teaching both authors (WEC and JMS) how to use another computer program for extracting fundamental frequency information (see Chapter 3) and for providing much-appreciated encouragement during the early growing pains of the project; Professors Jonathan Allen and Kenneth Stevens, for support and encouragement throughout; Dr. Louis Goldstein, Ursula Goldstein, and Dr. Howard Golub, for writing a computer program that transformed raw data into elegant graphs; Dr. Joan Bresnan, for unknowingly teaching JMS transformational syntax; Suzanne Boyce, Marcia Bush, Dr. George Clements, Martha Danly, and Sharon Hunnicutt, for asking JMS to help them learn Dr. Henke's computer program; Kittredge Cary, Dr. Ronald Cole, Jeanne Paccia-Cooper, Martha Danly, Farrid Dowla, Carmen Egido, Dr. Susan Ehrlich, Dr. Dennis Klatt, Dr. Mark Liberman, Julie Meister, Dr. Joseph Perkell, Janet Pierrehumbert, Dr. Herbert Simon, Howard Sherman, Dr. Saul Sternberg, and Dr. Ralph Vanderslice, for advice and discussion; Trude Huber, for help with the preparation of the languages table in Chapter 6; Trude Huber, Barbara Hussey, and Hedy Kodish, for much typing and retyping; and one hundred or so M.I.T. students and employees who served as volunteer speakers in the experiments.

Finally, we take this opportunity to thank each other, knowing that neither could have conducted the project alone. We enjoyed working together from the outset, in part because of the complementarity of our individual backgrounds and working styles. WEC came to the task from psycholinguistics; JMS, from electrical engineering. Our common denominator was an interest in acoustic phonetics, and from this basis we began the project, learning from each other as the work proceeded.

Each of us undertook our first acoustical studies of fundamental frequency during 1975. This work was largely unrelated to the concerns of the present treatise but enabled us to learn about computerized techniques for acoustical measurement. In 1976, we began this collaborative effort, concentrating much of the time between 1976 and 1979 on empirical research, with most of the writing carried out at Harvard University during the 1978–1979 academic year. In preparing the text, we benefited from oral presentations delivered earlier at the 95th Meeting of the Acoustical Society of America, the 14th Carnegie Symposium on Cognition, and in graduate seminars at both Harvard and M.I.T. Two early reports on the project were published as Cooper and Sorensen (1977) and Sorensen and Cooper (1980), listed in the References. The present treatment includes a few pages that have been revised from these earlier articles. However, nearly all the material included here is presented for the first time. Both research and writing were supported by grants from the National Institutes of Health (Grants NS 13028 and NS 15059).

Winter 1981

WILLIAM E. COOPER
JOHN M. SORENSEN

Contents

1. Introduction

While talking, we are generally unaware of the intricate movements of our vocal apparatus, just as we are unaware of our eye movements and other activities whose complexity is camouflaged by highly automated physiological systems. Only on rare occasions do we become conscious of these movements, despite the fact that they are governed for the most part by central commands originating from the brain.

During speech production, a number of structures are typically in motion at the same time. These structures include the lungs, the larynx, and supralaryngeal structures, such as the tongue, lips, and jaw. For present purposes, it is useful to focus on the intricate movements of the larynx. When speech is produced, the vocal folds in the larynx vibrate as air is pushed up from the lungs (see later for more details). The vocal folds separate as air is pushed upward; then, a Bernoulli pressure serves to pull the folds closer together (Sonesson, 1970). Momentarily, the folds come into contact and then begin to move apart again. This cyclic motion is repeated about 70 to 250 times per second during normal adult speech. The number of vibrations per second is termed the FUNDAMENTAL VOICE FREQUENCY and is symbolized as F_0.

Ordinarily, we are more likely to be aware of another person's eye movements than our own. Similarly, we are more readily aware of the perceptual correlate of F_0, known as VOICE PITCH. Most of us find it difficult to listen attentively to a monotone speaker, no matter how significant the content, while we typically find no trouble attending to the floral speech of radio broadcasters, no matter how INsignificant the content.

Variations in voice pitch seem to make listening easier, but these variations also reveal much about the speaker. Pitch provides clues to the speaker's age (Peterson and Barney, 1952), sex (Peterson and Barney, 1952), emotional state (e.g., Williams and Stevens, 1972), and personality (Scherer, 1974). In addition, it provides the listener with clues to the structure and meaning of the utterance (e.g., Fry, 1955; Streeter, 1978).

In this book, we focus primarily on the information that F_0 provides about the speaker's mental computations. Our chosen topic is commonly studied from a variety of vantage points—linguistic, psychological, physiological, and acoustic. Here, the topic is approached from a perspective that combines three of these four areas—the linguistic, the psychological, and the acoustic. The physiological activities are by no means neglected, but, as we shall see, research in this area has not proceeded to the stage where it can provide much information about the issues to be considered here.

Just as the study of the eye movements can be approached by examining the relationship between the movements themselves and environmental conditions that influence the viewer's central brain commands, so the study of F_0 is approached here by examining the relationship between F_0 itself and structural properties of utterances that influence the speaker's central control of vocal fold movement. In both cases, such inquiry can provide much information about the intricate mental operations that accompany some of our most intelligent yet routine behaviors. Since consequences of the behaviors themselves are readily observable and can be measured with a high degree of accuracy with modern computerized techniques, they provide a particularly clear window into thc cognitive workings of the brain. To the extent that the effort succeeds, the information obtained by such inquiry brings us closer to an understanding of human cognition and, additionally, often bears benefits to practical applications. In the case of F_0, such applications involve both communications engineering and speech pathology, as discussed in Chapter 6.

Previous Studies of Fundamental Frequency

Let us begin by briefly reviewing past work on F_0. These studies can be classified into three broad categories, dealing with linguistic structure, physiology, and acoustics. For the most part, prior research involving linguistics and acoustics will be discussed in later chapters in the context of our own studies. However, the anatomy and physiology of the larynx will be reviewed here in some detail, as a prerequisite to studies which deal less directly with these areas.

Linguistic Structure and F_0

The linguists' intuitions often provide guidelines for systematic experimentation, and, in the case of F_0, intuitions based on readily perceived differences in pitch permit one to make some initial hypotheses about the manner in which F_0 is influenced by linguistic structure. Informal observations suggest that F_0 is governed by three major grammatical properties of an utterance, including its mode (e.g., declarative vs. interrogative), stress pattern, and syntactic structure. While these properties can be distinguished in many cases, it should be noted that syntactic structure influences the stress pattern of an utterance to some extent, allowing for the possibility that F_0 may in some instances be influenced by syntactic structure indirectly via the mediating influence of stress. This possibility is shown in Figure 1.1. Much of the experimental work to be described in Chapters 2–4 deals with syntactic influences.

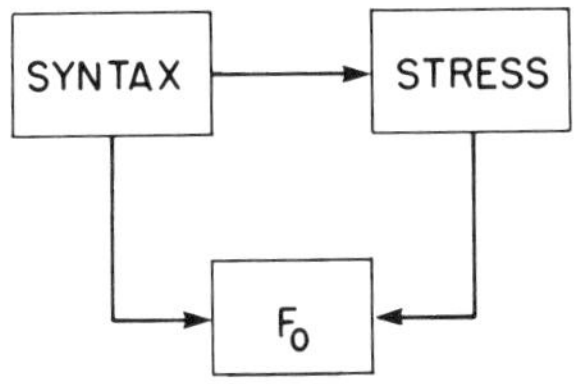

Figure 1.1. Possible routes of information flow among processing operations for syntax, stress, and F_0 during speech production.

Sentence Mode and Stress

In English and in many other languages, informal observations suggest that the F_0 pattern generally declines from the beginning to the end of a declarative utterance (Bolinger, 1964). For interrogative utterances, on the other hand, the last portion of the utterance is generally rising in F_0. This distinction between F_0 patterns in declarative vs. interrogative utterances is not observed universally, however. According to C. Grimes (personal communication), for example, one native Brazilian language includes gradually rising F_0 throughout declarative utterances. In Swedish, the F_0 rises near the end of declaratives, although a generally falling pattern prevails throughout the earlier portion of these utterances (Gårding, 1977a, 1979).

The gradual fall in F_0 throughout declaratives in English and many other languages has been termed DECLINATION, and we will investigate both the form and syntactic domain of this property in Chapter 2. As we shall see, the characteristics of declination are influenced in part by factors such as the syntactic structure as well as by physiological constraints on breathing. In fact, declination provides one of the most fertile grounds for studying the nature of the speaker's central commands in speech production, including the representation of syntactic units as well as the manner in which such units are processed during the speaker's planning and execution of speech.

The stress pattern of an utterance also exerts a substantial influence on F_0. Generally speaking, words receiving emphatic or contrastive stress exhibit a very high F_0 in comparison with neighboring words, although occasionally, emphasis is marked by an unusually low F_0 (e.g., Allen and O'Shaughnessy, 1974; Atkinson, 1976, 1978; Löfqvist, 1975; Maeda, 1976; O'Shaughnessy, 1976). In addition, stressed syllables typically receive higher F_0 than unstressed syllables when neither are marked by emphatic or contrastive stress. Aside from these broadbrush differences, subtle variations in stress pattern also produce influences on F_0, although in most of these cases the effects are small in magnitude and must be documented by direct acoustical measurements.

The influences of sentence mode and stress pattern have been studied extensively by British linguists, most notably in recent years by Halliday (1963, 1964, 1966, 1967) and Crystal (1969). In addition, recent linguistic studies have placed renewed emphasis on formulating a theory of metrical patterns in sentence contexts (e.g., Bresnan, 1971; Bolinger, 1972; Goldsmith, 1976; Leben, 1976; Liberman and Prince, 1977). For the most part, the influences on

F_0 studied throughout this book are orthogonal to the issues studied by these investigators, although ultimately the research of both types should be integrated into a model that captures the richness of F_0 variations as these occur in normal speaking. At present, however, it is useful to draw a sharp distinction between the influences of F_0 due to sentence mode and stress and the influence due to syntactic structure per se.

Syntax

The influence of syntactic structure on F_0 can be observed informally at a variety of major syntactic boundaries. These locations are typically marked by FALL–RISE patterns of F_0 and will be examined in detail in Chapter 3. Typically, the last stressed syllable preceding a major clause or phrase boundary is marked by an F_0 fall, while the first stressed syllable after the boundary is marked by an F_0 rise (Lea, 1973). These fall–rise patterns have been consistently observed in acoustical analyses of speech taken from spontaneous conversation as well as from the oral reading of paragraphs and isolated sentences (e.g., Lea, 1972, 1973). Occasionally, the F_0 fall, occurring just before the boundary, is immediately followed by a CONTINUATION RISE on the last stressed syllable preceding the boundary, as shown in Figure 1.2. In some languages (e.g., French), continuation rises appear commonly at such boundaries (Vaissière, 1975); in English, however, such rises appear less frequently (Maeda, 1976; Cooper and Sorensen, 1977). For this reason, in Chapter 3 we focus attention on English fall–rise patterns that encompass syllables on both sides of a syntactic boundary rather than on continuation rises that appear immediately preceding the boundary, though the latter are certainly worth examining in their own right (e.g., O'Shaughnessy, 1976, 1979).

Linguists traditionally relied on their perception of F_0 fall–rise patterns in conjunction with the perception of pauses to demarcate the syntactic structure of utterances (Trager and Smith, 1951). This effort was largely abandoned in the 1960s, however, when Lieberman (1965) showed this method of determining syntactic structures to be circular, by demonstrating that linguists based

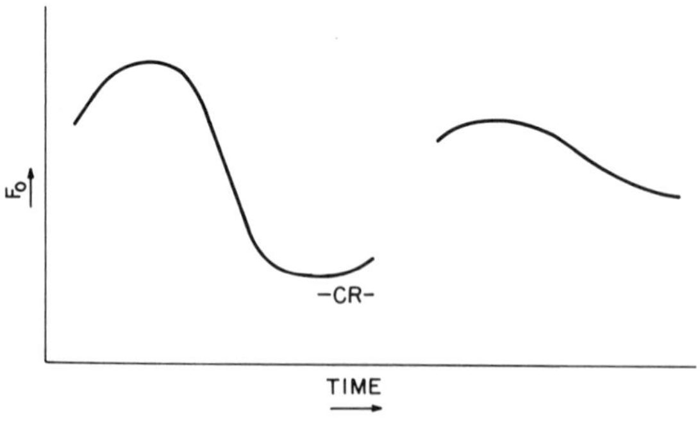

Figure 1.2. Schematic of fall–rise pattern showing a continuation rise (CR) on the word preceding the key boundary.

their perceptual judgments of intonation patterns on their intuitions about syntactic structure. Despite this finding, it is still true that major phrase and clause boundaries in normal speech are typically accompanied by fall–rise patterns of F_0, and, as we shall see in Chapter 3, acoustical measurements of such patterns provide a means of investigating the speaker's syntactic representation in more detail than has been possible by informal listening.

Speech Production and F_0

Experimental studies of the larynx and related structures have provided a useful way of examining the anatomical and physiological substrate of F_0 production. Because of the location of the larnyx, these studies involve intricate experimental techniques that are typically applied to only one or two speakers in a given study. One of the procedures involves taking high-speed motion pictures of the larynx during speech production after inserting a fiberoptic tube through the speaker's nasal passage (e.g., Sawashima and Hirose, 1968). Another method, sometimes used in combination with the fiberscope, consists of recording the electromyographic response of individual laryngeal muscles after inserting very small recording needles into the neck. Needless to say, both the fiberscope and electromyographic techniques involve some discomfort to the speaker, even when the insertions are carried out by the most competent of medical personnel. However, the discomfort seems to produce only a slight influence on the speaking patterns of veteran volunteer subjects.

With these experimental methods, in combination with acoustical measurements of F_0 during speech, much has been learned in recent years about the intricate workings of the larynx and the relation of laryngeal movements to the production of F_0. Some of this work will now be reviewed in the context of a particular issue that bears some relevance to one of the studies to be presented in Chapter 4, involving the manner in which laryngeal movements contribute to the distinction between the so-called "voiced" and "voiceless" plosives (e.g., [b] vs. [p]).

Anatomy and Physiology of the Larynx

Numerous accounts of laryngeal anatomy have been presented, but one of the most useful for present purposes is the review provided by Sonesson (1970). More recent studies (e.g., Hirano, 1974) have provided additional information, and both traditional and recent observations will be included in the following characterization.

Viewed generally, the larynx is composed of cartilages connected by joints and membranes, together with several small muscles that move the joints. From the standpoint of speech production, the main parts of the larynx include the vocal folds and the glottis. The vocal folds are composed of the vocalis muscle, the conus elasticus (a funnel-like connective tissue membrane that

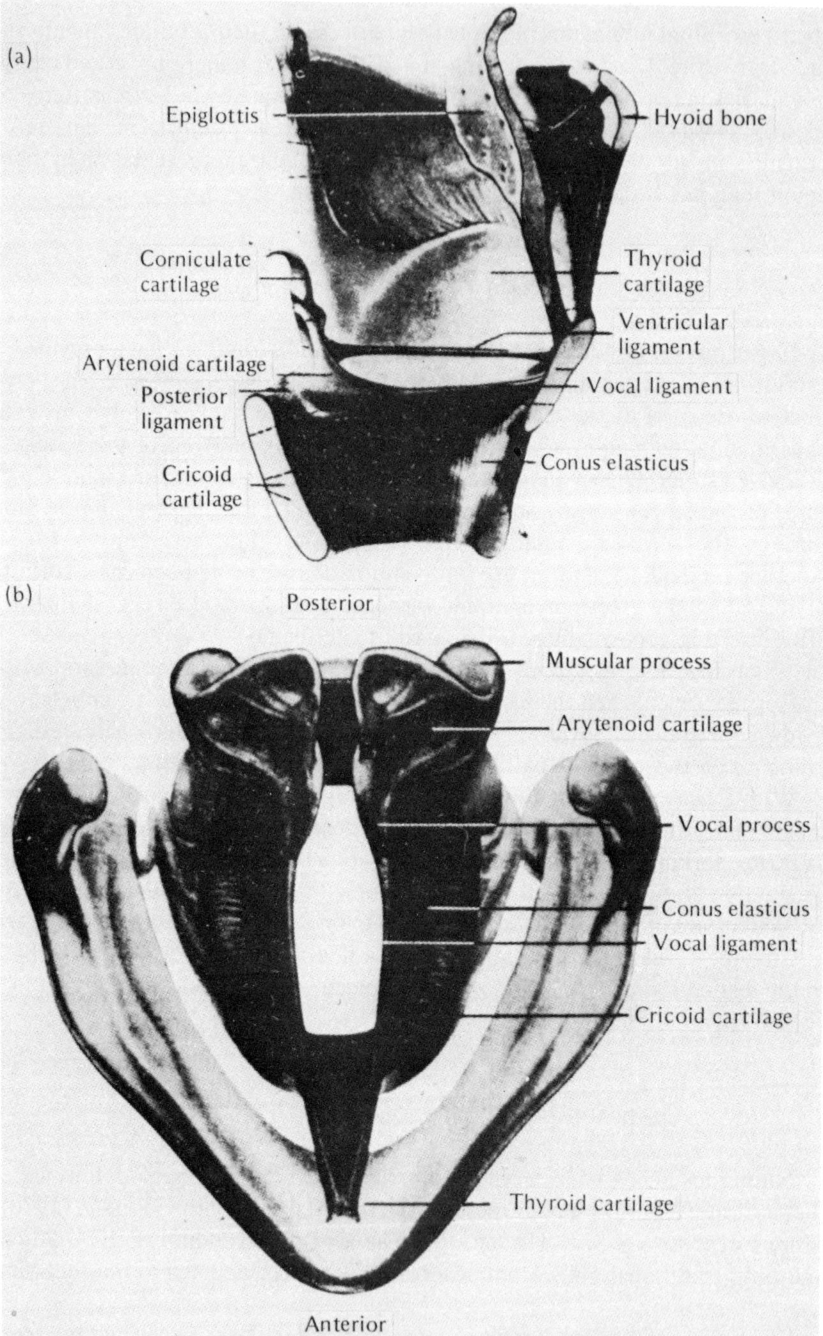

Figure 1.3. The cartilages, ligaments, and membranes of the larynx as viewed from the inner side (a) and from above (b). Reprinted with permission from Pernkopf (1952).

stretches up from the cricoid cartilage), and the vocal ligament. The cavity between the vocal folds is the glottis, which forms a triangular slit during the rest position of respiration. Changes in the state of the vocal folds and the glottis are controlled in large part by other components of the larynx, as described below.

Cartilages

The larynx includes three unpaired and three paired cartilages. The unpaired cartilages include the thyroid, cricoid, and epiglottis; the paired ones include the arytenoid, corniculate, and cuneiforme. Largest of the cartilages is the thyroid, a V-shaped structure whose open angle is turned backward, and whose anterior border corresponds to what we know as the Adam's apple. The general shape and position of the thyroid, along with the other cartilages, are shown in Figure 1.3.

Joints

The joints of the larynx, whose movements regulate the tension and position of the vocal folds, include two paired joints, the cricothyroid and cricoarytenoid. In the cricothyroid joint, the cricoid cartilage undergoes a rotational movement around an axis of horizontal transverse rotation, this axis being common to the pair of joints (see Figure 1.4). Two types of rotation are of interest. If the cricoid arch is raised during the rotation toward the anterior of the thyroid cartilage, then the cricoid laminae (two parts of the cricoid corresponding to the legs of the letter "V") are tilted backward. This tilting results in a length-

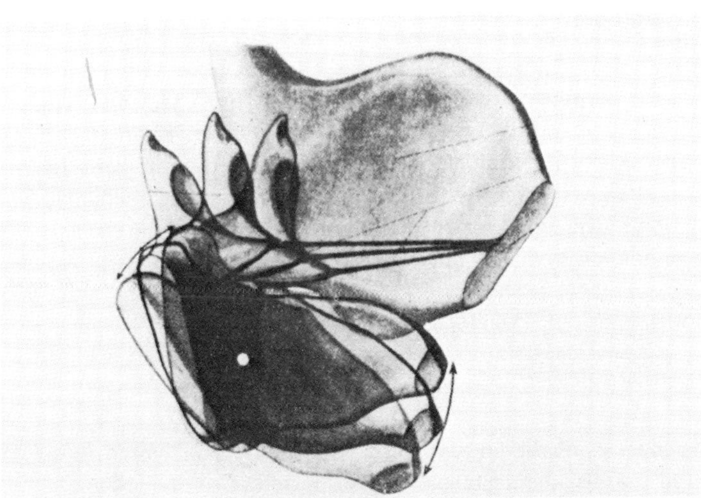

Figure 1.4. The rotation movement in the cricothyroid joint in the larynx. The white spot indicates the axis of rotation. Reprinted with permission from Pernkopf (1952).

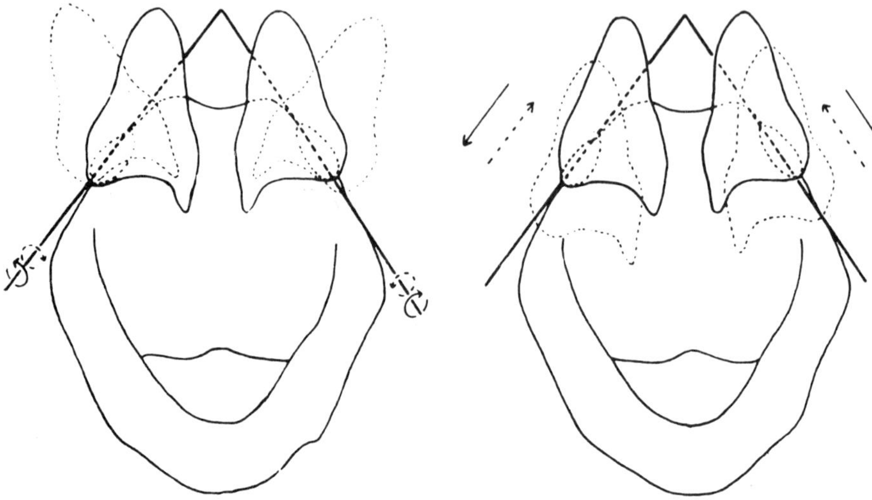

Figure 1.5. The two types of movement in the cricoarytenoid joint. Reprinted with permission from Sonesson (1970).

ening and stiffening of the vocal folds, which themselves run between the thyroid and the arytenoid cartilages. Alternatively, if the cricoid arch is lowered, then the cricoid laminae are tilted forward, resulting in a shortening and slackening of the folds. The movements of the cricothyroid joints thus determine in part the degree of isotonic tension of the vocalis muscle within the vocal folds.

Consider now the movements of the cricoarytenoid joint (following Sonesson, we shall henceforth refer to paired laryngeal structures in the singular form), which is enclosed by a capsule-like ligament and by a posterior ligament connecting the arytenoid and cricoid cartilages. This joint is formed by a concave part of the underside of the arytenoid cartilage and by a convex part of the top of the cricoid cartilage lamina. The axis of movement of the cricoarytenoid joint permits two types of movement by the arytenoid cartilage, including rotation and translation (see Figure 1.5). During rotation, the VOCAL PROCESS (the base of the arytenoid cartilage connected to the vocal folds) is either raised and moved laterally, or is lowered and moved medially. During the translation movement, the arytenoid cartilage moves along the axis of rotation, being either lowered and moved forwards, or raised and moved backwards. These movements of the arytenoid displace the vocal folds along a superior–inferior dimension as well as laterally and medially.

Muscles

The muscles of the larynx consist of a number of paired muscles and one unpaired muscle, the arytenoid. These muscles are divided anatomically into two major types, the intrinsic muscles, which regulate the mutual position of the cartilages within the larynx, and the extrinsic muscles, connecting parts of the larynx to other skeletal structures.

The intrinsic muscles (see Figure 1.6) operate primarily to control the opening and closing of the glottis as well as the degree of tension of the vocal folds. Opening of the glottis is accomplished by activity within the posterior cricoarytenoid muscle, which rotates the vocal process upward and laterally, leaving a space between the arytenoid cartilages.

Each of the remaining intrinsic muscles serves to close the glottis. This functional emphasis on the closing of the glottis no doubt reflects the importance of the larynx in protecting the lungs (e.g., Lindqvist, 1970). As we shall see, however, regulating the static opening of the glottis is also of major importance for producing distinctions among the plosive consonants during speech.

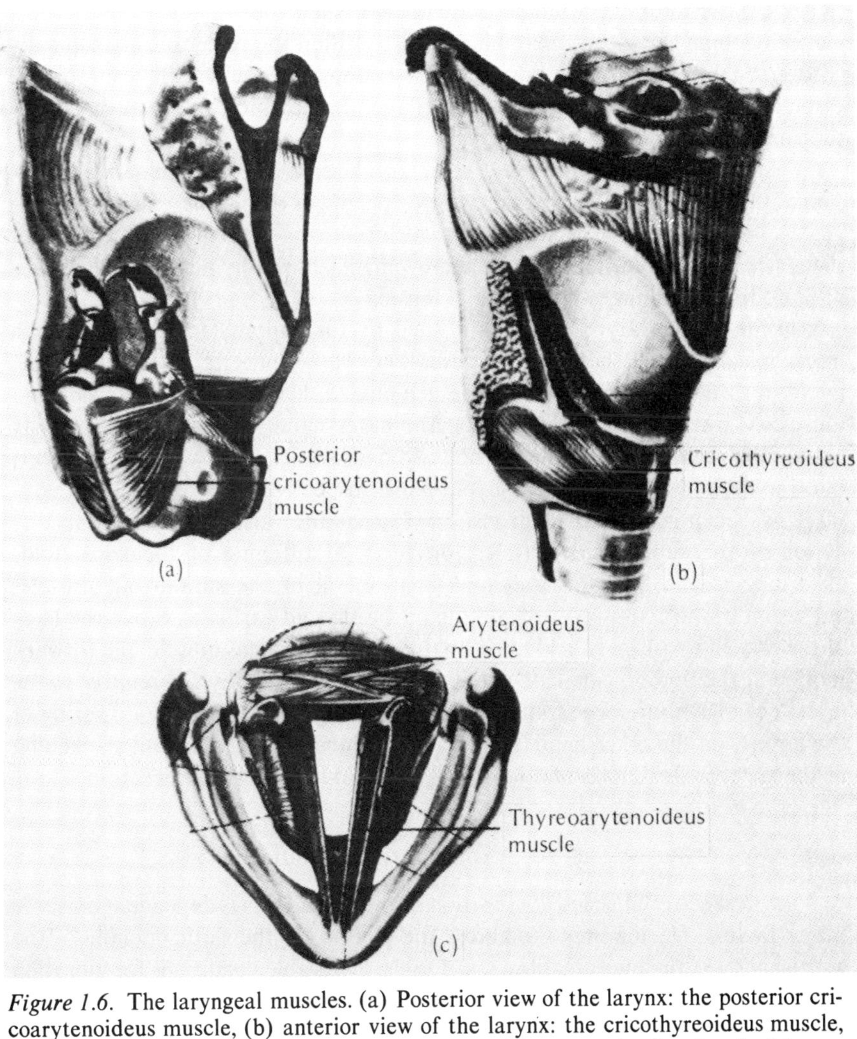

Figure 1.6. The laryngeal muscles. (a) Posterior view of the larynx: the posterior cricoarytenoideus muscle, (b) anterior view of the larynx: the cricothyreoideus muscle, (c) the larynx seen from above: the thyreoarytenoideus muscle. Reprinted with permission from Pernkopf (1952).

The thyroarytenoid muscle consists of two parts, a medial portion, known as the vocalis muscle, and a lateral portion. Traditionally, the vocalis is believed to regulate the tension of the vocal folds isometrically, whereas the lateral portion of the thyroarytenoid serves primarily to close the glottis. In the case of the vocalis, however, Hirano (1974) has pointed out that the activity within this muscle primarily affects the inner part of the vocal folds only, including the conus elasticus but excluding the mucous membrane that covers the vocal folds. Hirano (1974) postulates that, during vibration of the vocal folds, the relations between the body and cover of the folds are controlled by the combined activity of the vocalis and cricothyroid muscles (see also Fujimura, 1977).

Before leaving the discussion of intrinsic laryngeal muscles, we should note that the posterior cricoarytenoid muscle, mentioned above as the primary muscle for opening the glottis, may also serve to stiffen the vocal folds isotonically (Stevens, 1977). By contracting this muscle, the arytenoid cartilages would be displaced posteriorly on the cricoid cartilage, thereby increasing the tension of the vocal folds.

Whereas the intrinsic muscles of the larynx serve primarily to vary the tension of the vocal folds as well as the static opening of the glottis, extrinsic laryngeal muscles are primarily responsible for producing vertical displacements of the larynx as a whole. These vertical movements, however, can themselves produce additional changes in vocal fold tension.

Raising of the larynx is accomplished by the suprahyoid muscles, which insert into the hyoid bone (see Figure 1.7). These muscles diverge from the hyoid in three directions to form the geniohyoid, hyoglossus, and sternohyoid muscles. Lowering of the larynx, on the other hand, is achieved by activity within the sternothyroid and sternohyoid muscles. These two muscles insert into the thyroid lamina and the hyoid bone, respectively.

It has been suggested (Stevens, 1977) that the raising of the larynx, by action of the suprahyoid muscles, produces a stiffening of the vocal folds according to the following scheme. Contraction of the suprahyoid muscles causes an upward and forward motion of the thyroid cartilage. Since the back of the cricoid cartilage is held fixed by the cricopharyngeus muscle, the forward motion of the thyroid cartilage results in a sliding of the thyroid relative to the cricoid cartilage, thereby stretching and stiffening the vocal folds. Lowering of the larynx, on the other hand, appears to produce a thickening and slackening of the vocal folds (Ohala and Hirose, 1970; Kakita and Hiki, 1972).

The Vocal Folds during Vibration

The presence or absence of vibration of the vocal folds is dependent on three major factors: (1) the pressure across the glottis; (2) the static opening of the glottis; and (3) the tension of the vocal folds. Favorable conditions for vibration of the vocal folds include a sufficiently high transglottal pressure, a nearly closed glottis, and slackened vocal folds (Halle and Stevens, 1971).

Much is known about the detailed aerodynamic and mechanical properties

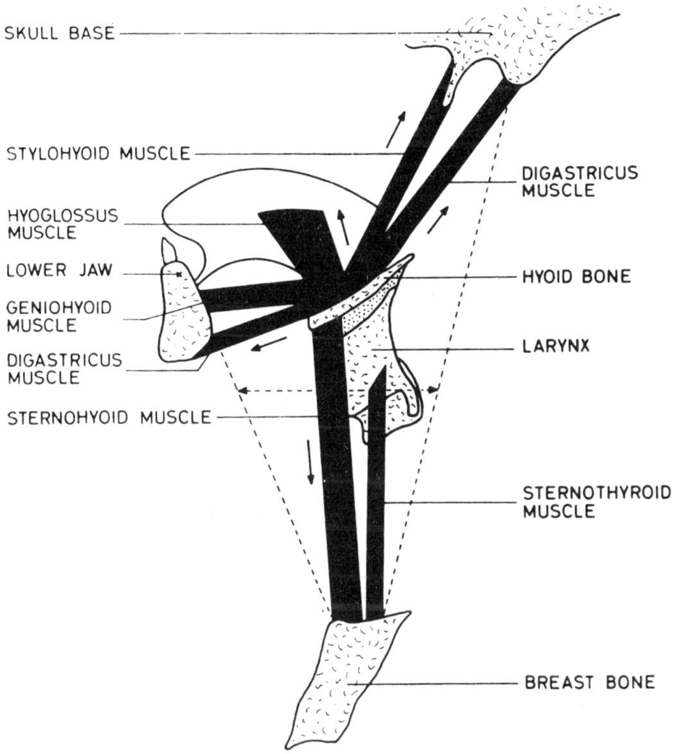

Figure 1.7. The extrinsic laryngeal muscles or supra- and infrahyoidal muscles. Reprinted with permission from Sonesson (1970).

of vocal fold vibration (e.g., van den Berg, 1970; Ishizaka and Matsudaira, 1968; Baer, 1975; Titze, 1976). Although these properties do not play a crucial role in plosive distinctions, it is worthwhile to mention just one of the detailed characteristics of the ongoing vibration pattern. Ishizaka and Matsudaira (1968) have observed that a phase difference exists between the movement of the upper and lower edges of the vocal folds during vibration and that this phase difference is essential for maintaining oscillation. The lag in phase allows for the necessary transfer of energy from the steady airflow across the glottis to the vibrating vocal folds. A two-mass model of the vocal folds has been devised which incorporates this phase lag feature (Ishizaka and Flanagan, 1972). With this model, it has been possible to duplicate a number of the characteristics of human vocal fold vibration, thereby permitting quantitative investigation of laryngeal parameters. A three-dimension model developed by Titze (1976) provides an even more realistic means of simulating the vocal fold movements.

While a number of properties of ongoing vocal fold vibration have been mapped out, comparatively little is known about the factors involved in the INITIATION of such vibration, and it is this event which is of primary concern for understanding how distinctions among the plosive consonants are produced.

Using the background in anatomy and basic physiology of the larynx, we can now begin to consider this issue.

Laryngeal Features for Plosive Consonants

Two Major Alternatives

For syllable-initial plosives in English, the distinction between the plosives [b]-[p^h], [d]-[t^h], and [g]-[k^h] can be viewed in a number of ways, suggested by differences observable from a speaker's acoustic output as well as differences noted by recording muscle potentials or photographing the larynx. For one thing, there exists a clear difference in the timing of the onset of vocal fold vibration for these plosives, relative to the release of the obstruction in the supralaryngeal tract. For the voiced plosives [b], [d], and [g], vocal fold vibration begins either before or shortly after the release of closure (in most instances, there is a short lag in voicing onset of less than 20 msec), whereas for the voiceless plosives [p^h], [t^h], and [k^h], there is a much longer lag in the onset of vocal fold vibration, typically longer than 40 msec. In a series of studies, Lisker and Abramson (1964, 1967, 1971) have investigated this timing dimension, termed voice onset time (VOT). Lisker and Abramson (1971) have reached the conclusion that VOT is the basic determinant for the differences between the voiced and voiceless plosives observed in natural languages. In physiological terms, they hypothesize that a centralized timing control mechanism exists for regulating the onset of vocal fold vibration relative to the release of the supralaryngeal closure. This view has grown out of extensive cross-language studies in which they have shown that the VOT relation sufficiently distinguishes the plosive consonants within a number of different languages.

The timing account of plosive distinctions would stand unchallenged if no other observable differences existed in the acoustic output of voiced vs. voiceless plosives, but, as Lisker and Abramson note, such additional differences do exist. These differences open up the possibility that the timing relation of VOT is not under direct physiological control but is rather a consequence of other, independently controllable parameters, whose timing is controlled relative to other events. This position is adopted by Halle and Stevens (1971), who attempt to account for the variety of plosive distinctions and VOT relations observed in natural languages in terms of two independent laryngeal parameters, static opening of the glottis and vocal fold tension.

With these two parameters, Halle and Stevens propose 4 binary-valued laryngeal features. These include $\pm$ spread glottis, $\pm$ constricted glottis, $\pm$ stiff vocal folds, and $\pm$ slack vocal folds, with the provision that the feature combinations [+ spread, + constricted] and [+ stiff, + slack] cannot occur. When the glottis is spread and the vocal folds are stiff, the vibration of the vocal folds will be maximally inhibited. On the other hand, when the glottis is approximated and the vocal folds are slack, the conditions for vibration will be optimally favorable. For a given glottal opening and transglottal pressure, vibration of the vocal folds will begin only if the folds are sufficiently slack, while for a given vocal fold stiffness and transglottal pressure, vibration will begin only if the glottis is sufficiently constricted.

Both the Lisker–Abramson and the Halle–Stevens hypotheses are consistent with a large portion of the data on plosives in various languages. Unfortunately, no critical test has been conducted which clearly differentiates their merits. We can, however, consider an issue which bears on one aspect of these proposals. If the timing relation for the voiceless plosives between the onset of vocal fold vibration and the release of closure is under CLOSED LOOP CONTROL (i.e., information about the release of closure is fed back to a central controller which then initiates glottal adduction and related activities leading to vocal fold vibration), then, for a given plosive type, there should be less variability in the VOT values for that plosive than in the values of laryngeal parameters such as glottal opening and vocal fold tension. This closed loop hypothesis was not considered in earlier work, but, if supported, would certainly provide a strong peg of evidence favoring the general view of Lisker and Abramson.[1] Unfortunately, there is no way of telling what "less variability" means in this case, since we have no means of deciding on a proper comparison of the variability of VOT values and the variability of glottal opening or vocal fold tension measures.

A good test of this hypothesis would involve a direct comparison of the static glottal opening at the release of oral closure and the time course of glottal adduction, in conjunction with measurements of voice onset time. In a study of Korean plosives, Kagaya (1974) has performed such a test.

The sound system of Korean is unusual in that it contains a three-way distinction among voiceless plosives for a given place of articulation. Type 1 voiceless plosives [P, T, K] are typically referred to by phoneticians as unaspirated and tense; Type 2 plosives [p, t, k], as slightly aspirated and lax; and Type 3 plosives [p^h, t^h, k^h], as heavily aspirated and tense. Kagaya (1974) investigated the glottal width functions for these plosives by means of high-speed laryngoscopy, using a fiberoptic bundle inserted through the nostril. With this technique, pictures of the glottal opening can be taken at speeds up to 60 frames/sec (Sawashima and Hirose, 1968; more recent improvements permit even higher speeds). Kagaya's pictures were taken at a rate of 50 frames/sec for one subject and at 24 frames/sec for a second subject.

Considering the pooled data from a variety of utterances for an individual speaker, Kagaya found an almost constant rate of glottal close-down for both Type 2 and Type 3 voiceless plosives after the release of oral closure [for Type 1 voiceless plosives, glottal adduction was completed by the time of the oral release (see Kagaya, 1974, Figure 2)]. Prior to the oral release, however, the glottal width functions for the Type 2 and Type 3 plosives were more variable

[1]Lisker and Abramson (1971) consider the possibility that VOT is generally under the speaker's control for achieving the distinction between voiced and voiceless plosives. However, they do not make the stronger claim that VOT is under closed loop control in individual utterances.

both within and across types. After the release of closure, the rate of glottal adduction was nearly constant and equal for the two types.

These findings suggest two important principles: (1) the rate of glottal close-down after the release of closure may be independent of changes in glottal width that occur just prior to the release; and (2) the close-down rate after the release is fixed and not under closed loop physiological control. However, the evidence against the closed loop hypothesis would be compelling only if the obtained rate of glottal close-down were perfectly constant, and Kagaya's data do not meet this strict requirement. As long as the rate is not precisely constant, it can be said that this rate is indeed under closed loop control but that the variations produced as a result of this active control are quite small.

A number of other studies have been conducted to examine aspects of the hypothesis of Halle and Stevens (1971) concerning their feature [±stiff] (e.g., Hardcastle, 1973; Hirose and Gay, 1972; Maeda, 1975). However, the data obtained from these studies are by no means in agreement (see, for example, Hirose and Gay, 1972, vs. Maeda, 1975). In large part, the relative merits of the Halle–Stevens vs. Lisker–Abramson proposals still remain uncertain (see also Ladefoged, 1973; Stevens, 1977).

Vertical Displacements of the Larynx

Hudgins and Stetson (1935) observed that the production of voiced plosives was accompanied by a lowering of the larynx as a whole, whereas the production of voiceless plosives was accompanied by larynx raising. A convenient method has been developed to measure these relative displacements of the larynx during normal speech (Kakita and Hiki, 1972). A small mirror is attached to the skin near the thyroid cartilage; vertical movements of the thyroid (or cricoid) cartilage tilt this mirror, and a light beam reflected from the mirror is converted into a voltage by means of a variable-transparent optical mask and photocell. In this way, a continuous record of the relative position of the thyroid cartilage can be obtained (see also Ewan and Krones, 1973, 1974).

Using this technique, Kakita and Hiki (1972, 1974, 1976) have shown that, in the production of Japanese word accent, an upward displacement of the larynx is correlated with a rise in fundamental frequency, whereas a downward displacement is correlated with a falling F_0. It thus appears possible that the vertical displacements of the larynx play a role in producing tension on the vocal folds. Electromyographic records (Kakita and Hiki, 1972) show that an increase in the activity of the thyrohyoid muscle accompanies the raised larynx and that an increase in the activity of the sternohyoid and sternothyroid muscles accompanies a lowered larynx.

For the plosive consonants, X-ray motion pictures taken for Korean plosives by Kim (1965) show a correlation between raising the larynx and a rise in F_0 of a following vowel, although a similar study by Kagaya (1971) has shown a correlation between larynx *lowering* and a rise in F_0. Whatever the explanation for this discrepancy, it is clear that monitoring of the extrinsic laryngeal muscles should accompany studies that are designed to test the presence of the tensity feature for plosives, since it appears that vocal fold tension can be altered not only by action of the vocalis and other intrinsic laryngeal muscles, but also by action of the extrinsic laryngeal muscles (see also Sawashima, Kakita, and Hiki, 1973).

More recently, Fujimura (1977) has developed a stereofiberscope in order to measure the vertical displacement of the larynx by direct visual inspection. By inserting a glass-fiber cable through each of the two nostrils, it is possible to obtain two separate images of the larynx, and the corresponding positions from each image can be used by a computer program to derive information about three-dimensional movements of the larynx. With the advent of this new measurement procedure, it should be possible to obtain more information about the role of larynx height (see Fujimura, 1977).

The Relative Timing of Laryngeal Gestures

In the preceding discussion, we have seen that the larynx undergoes a number of different changes during the production of plosive consonants, including changes in glottal width, vocal fold tension, and vertical position of the larynx. But how is the temporal sequencing of these changes controlled? To find out, one must turn to studies of the laryngeal muscles in which simultaneous recording of both the intrinsic and extrinsic laryngeal muscles has been obtained. Unfortunately, such work is scarce at present, but some evidence does exist from the work of Kakita and Hiki (1974). Before turning to this evidence, it may be useful to consider evidence relating to temporal sequencing of activity within the intrinsic laryngeal muscle group.

Hirose and Gay (1972) have found a number of instances in which the muscle activity as measured by EMG consisted of a reciprocal pattern for two types of muscle. For example, they found that the interarytenoid muscle showed a pattern of activity which was approximately reciprocal to that of the posterior cricoarytenoid, with an increase in the activity of one of these muscles being accompanied by a simultaneous decrease in the activity of the other. The reciprocal pattern of activity for these muscles occurred for the production of both voiced and voiceless plosives in English, with the interarytenoid muscle playing the role of an adductor of the vocal folds, and the posterior cricoarytenoid playing the role of an abductor. For the latter muscle, the peak in activity as measured by EMG occurred well before the release of oral closure, by as much as 165 msec in the case of a poststressed [p].

Hirose and Gay found in addition that the timing of posterior cricoarytenoid activity with respect to supralaryngeal activity differed depending on the context of the voiceless plosive. For poststressed plosives, the time at which the posterior cricoarytenoid activity reached a minimum occurred before or simultaneous with the release of oral closure. However, for PREstressed plosives, the activity of this muscle reached a minimum only after the release of closure had occurred. This finding suggested that the glottis is partially abducted at the moment of oral release in the case of prestressed voiceless plosives, consistent with the lengthy voice onset time observed for such consonants (Lisker and Abramson, 1967).

While the data of Hirose and Gay's study provide useful information about the physiology of the intrinsic laryngeal muscle group, such data do not tell us about the relative timing of activity in the intrinsic muscles with respect to the extrinsic muscles. Kakita and Hiki (1974) have performed simultaneous recording of both intrinsic and extrinsic laryngeal muscles, although their study was performed during the production of distinctions in Japanese word accent in the utterance [namanama] rather that during plosive production.

Kakita and Hiki found that the activity of the extrinsic muscles, in particular the thyrohyoid, sternothyroid, and sternohyoid, lagged slightly behind the activity of those instrinsic muscles which play a major role in producing changes in vocal fold tension, including the vocalis and cricothyroid (see also Sawashima, Kakita and Hiki, 1973). Given these data, however, there is no way of determining whether the timing of the activity of these two groups of muscles is regulated by a single control mechanism or whether the timing of each muscle group is regulated quasi-independently of the other. Although the former possibility is more likely, it would be of interest to test the possibility of quasi-independent timing by comparing individual utterances and determining the extent to which changes in the timing of the intrinsic muscle activity (relative to oral release) correlate with the timing of extrinsic muscle activity; the greater the correlation, the more probable that the timing of these two muscles are controlled by a unitary mechanism.

Conclusion

Whereas the nature of ongoing vibration of the vocal folds is relatively well understood, conflicting views have been proposed for explaining how such vibration is initiated following word-initial voiced and voiceless plosives. In this brief review, we have tried to point out the salient features of these views and show how physiological data can be brought to bear on these and related aspects of the problem. For present purposes, this background in laryngeal anatomy and physiology serves as a prerequisite for understanding the acoustical consequences of vocal fold vibration to be discussed throughout the remainder of this book. For the most part, the acoustical studies on F_0 will involve the measurement of F_0 during ongoing vocal fold vibration, typically during the production of vowel segments within words spoken in sentence contexts. However, in Chapter 4 we shall present a study that involves the measurement of F_0 as vocal fold vibration is initiated following the production of voiced and voiceless plosives.

In addition to the studies of laryngeal activity, studies of subglottal pressure (i.e., the pressure below the laryngeal region) have also been useful in characterizing the production of F_0. Originally, it was believed that differences in subglottal pressure accounted for most of the variations in F_0 that accompany words of different stress (Lieberman, 1967). However, more precise measurements have revealed that the variations in F_0 attributed to stress are primarily produced by variations in the tension of the laryngeal muscles themselves (Hixon, Klatt, and Mead, 1971). Collier (1975) and Maeda (1976) have noted, however, that differences in subglottal pressure do seem to account for a portion of the F_0 declination that accompanies declarative sentences. The relationship between subglottal pressure and F_0 declination will need to be examined further in light of the acoustical studies to be presented in Chapter 2.

Acoustics and F_0

Acoustical studies have shown that F_0 variations are influenced by a variety of grammatical factors, including the inherent phonetic segments on which F_0 is superimposed, the phonetic structure of neighboring segments, stress, and syntactic structure. All these factors combine to influence the F_0 values contained in any particular utterance. When attempting to examine one of these factors in detail, it is thus necessary to control or neutralize the influence of the others. Many attempts to provide such control will be presented in the studies of Chapters 2–4.

The F_0 value is determined to a small extent by the inherent phonetic segment. High vowels such as /i/ and /u/ are accompanied by slightly higher F_0 values than low vowels such as /a/ or /æ/, other factors being equal (Peterson and Barney, 1952). Lisker (1974) has argued that the differences in F_0 for various vowels represent a consequence of the vocal fold tension which typically accompanies these vowels. The differences in F_0 for high vs. low vowels is small, however, representing only about 5% of the F_0 value.

The F_0 value is also determined by the structure of neighboring segments, as expected on the basis of the earlier discussion of laryngeal features. For example, the F_0 pattern for a vowel following a voiceless plosive typically exhibits a falling pattern, whereas the F_0 for a vowel following a voiced plosive typically exhibits a rising trend (House and Fairbanks, 1953). The difference here may be attributed to the tension on the vocal folds accompanying the production of the plosives themselves. As noted earlier, for voiceless plosives, the folds are relatively tense, whereas for voiced plosives, the folds are lax. Thus, at the beginning of the vowel, the F_0 is higher after a voiceless than after a voiced plosive. The F_0 glides downward to a steady state after a voiceless plosive and glides upward to the same state after a voiced one. As we shall see in Chapter 4, this effect of phonetic environment might operate across word boundaries to some extent, where the plosive is at the end of one word and the vowel begins the next. This cross-word conditioning effect is particularly interesting because of the possibility that it may be blocked at certain major syntactic boundaries.

The influence of stress pattern on F_0 has been studied in a variety of acoustical studies (e.g., Lea 1972; Allen and O'Shaughnessy, 1974; O'Shaughnessy, 1975; Liberman, 1975). Generally speaking, these studies not only document quite obvious differences in F_0 as a function of major variations in stress, but they are also concerned with effects of stress that often go unnoticed during informal listening.

Finally, the influence of syntactic structure on F_0 has been studied acoustically by Lea (1972, 1973), O'Shaughnessy (1976), and Maeda (1976), among others. These studies have shown that, as expected on the basis of linguistic intuition, major syntactic boundaries are often accompanied by fall–rise patterns of F_0. However, these studies generally include only a few subjects per experiment with relatively loosely controlled sentence materials, making it difficult to make general conclusions about the role of syntactic influences. In Chapter 3, we shall present a systematic study of fall–rise patterns that attempts to assess syntactic influences in a more tightly controlled setting with a relatively large number of speakers.

The Present Study

Phenomena

Of the various influences on F_0 that have been reviewed thus far, syntactic influences comprise the primary focus of our own research. On the basis of earlier studies, we can identify at least one attribute of F_0 that appears to be influenced by syntax, namely, fall–rise patterns, which typically accompany major syntactic boundaries. In addition to these patterns, we add two types of potential syntactic influence on F_0. These include influences on the domain of F_0 declination and the blocking of cross-word phonetic conditioning effects on F_0. The three phenomena of fall–rise patterns, F_0 declination, and cross-word phonetic conditioning effects on F_0 will be studied here in order to determine how the speaker represents and processes syntactic information during speech production. Each of these three phenomena is illustrated in the F_0 pattern for the utterance shown schematically in Figure 1.8. While this work is aimed largely at discovering properties of the speaker's syntactic code, it will be necessary to examine more closely the acoustic attributes of these three F_0 phenomena themselves. Consequently, the study should provide information not only for cognitive psychologists and linguists, but also for communication engineers and others who are interested in a variety of applications.

Issues

Two major questions can be asked about the speaker's mental computations involving syntactic structures: (1) What are the structural units represented during speech production? and (2) How are these units processed during the planning and execution of speech itself? In previous work, these questions have been studied by examining another speech variable—the timing of segments and pauses (e.g., Cooper and Paccia-Cooper, 1980). Previous linguistic obser-

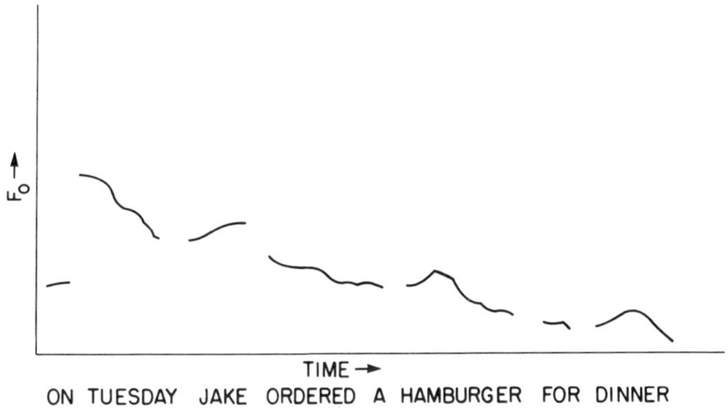

Figure 1.8. Schematic F_0 pattern for the utterance *On Tuesday Jake ordered a hamburger for dinner,* illustrating the phenomena of declination, fall–rise patterns, and cross-word phonetic conditioning. Declination can be observed throughout the course of the utterance; a typical fall–rise pattern appears at the syntactic boundary between the words *Tuesday* and *Jake;* cross-word phonetic conditioning might occur between the final /k/ segment of *Jake* and the beginning vowel segment of "ordered" inasmuch as the F_0 values at the beginning of the verb are somewhat higher than if the preceding word had ended in a voiced consonant (see Chapter 4).

vations had informed us that both timing and F_0 might be appropriate dependent variables to measure in seeking answers to questions about the speaker's coding of syntactic structures. Speech timing was chosen first because it is a simpler variable to measure and to interpret. Segments and pauses can be either longer or shorter in duration, but no other relative variations in timing are permitted with this unidimensional variable. Despite its simplicity, speech timing provided a great deal of new evidence about the speaker's coding of syntactic structures. Because of the greater complexity of F_0 patterns, however, it is quite possible that F_0 will provide an even richer source of information about the speaker's internal code.

Methods

Earlier linguistic and acoustic observations have indicated that syntactic structures do influence F_0, yet the nature of this influence remains largely unexplored. The reason for this shortcoming is that previous studies have typically emphasized the linguistic and/or acoustic aspects of the problem without considering the psychological issues involved, both in terms of methodology and the interpretation of results. The testing of syntactic influences on F_0 has utilized a few number of speakers and sentence materials that were often uncontrolled for extrasyntactic influences known to affect F_0. These shortcomings have been largely overcome in a series of studies on syntactic influences of speech timing, and these studies will serve as a model for the present investigation.

Sentence-Reading Procedure

The timing and F_0 studies both utilize a sentence-reading procedure, whereby speakers are asked to read sentences that have been carefully constructed to include particular variations in syntactic structure. This method allows one to combine careful control over extrasyntactic influences on F_0 with relatively natural utterances.

In constructing the sentence materials to be used in these studies, one tries to create strings that are quite typical of conversational English and yet permit an adequate test of the particular issue under investigation. This dual concern marks a departure from earlier experimental studies of F_0 in which the sentences were typically quite awkward (e.g., Atkinson, 1976; Lea, 1974; Olive, 1975). From our standpoint, the face validity of the experimental task depends on a joint consideration of both naturalness and tight experimental control. Though these two factors are typically in conflict with one another, one can attempt to optimize them in constructing test sentences.

Despite the advantages of the sentence-reading procedure in terms of experimental control, it would still be ideal to study F_0 contours using a corpus of spontaneously generated speech. To date, this goal is impractical for all but the most broadbrush studies of F_0 features, since the data analysis cannot untangle the various influences on F_0 in a systematic way. However, two of the major F_0 phenomena to be studied here—F_0 declination and fall–rise patterns—do appear reliably in typical conversation (M. Liberman, personal communication, May, 1978) as well as in unpracticed reading of a lengthy passage of text (Stevens, Henke, and Sorensen, 1976). It seems reasonable to assume that detailed studies of declination and fall–rise patterns with a sentence-reading procedure should at least provide a useful foundation for constraining subsequent studies devoted to spontaneous speech. In addition, the studies with sentence-reading provide direct input to a variety of applications to be discussed in Chapter 6.

The sentences in the experiments themselves contain one or more key words to be measured for F_0. These key words are typically chosen so as to be easily segmentable from the acoustic waveform.

In a given experimental session lasting about 50 min, an individual speaker is presented from about seven to ten short lists of sentences. Each list begins and ends with a filler sentence that is similar in structure and meaning to the actual test sentences so as to camouflage the purpose of the experiment. Fillers are also typically interspersed within the list. The order of presentation of the sentences in the lists is pseudorandomized and also counterbalanced for half of a group of about 10 speakers. Each sentence is typed on a single line without any underlining or punctuation except for a final period. Although this procedure was adopted throughout the experiments, test sentences will typically be presented in this text with the key words underlined for the reader's convenience.

The speakers include area college students and employees who serve as paid volunteers. The subjects are always native speakers of American English with no impairments in their speech or hearing.

These speakers are tested individually in a large sound-insulated chamber. The speech of the subjects is recorded via a high-quality voice microphone (Altec 684A) onto a professional tape recorder (Presto A908). The tape recorder is located in an adjacent room and is under the control of the experimenter. All utterances are recorded at a speed of 7½ inches/sec. Between trials, the experimenter and subject can communicate with each other, either visually through a large window separating the two rooms or verbally upon the experimenter's entrance to the sound-insulated chamber. The subject is asked to consider each sentence individually. For each sentence, the speaker first reads the sentence silently to gain an understanding of the meaning and then reads it aloud once for practice and to enable the experimenter to check recording levels. The speaker then reads the sentence twice for recording. For all readings, he is instructed to speak in a normal conversational voice and to avoid reading word-by-word as in unpracticed reading. On each trial, the experimenter cues the speaker after listening to the practice utterance unless this utterance is determined unsuitable for reasons of contrastive or emphatic stress, as noted below. The experimenter starts the tape recorder just prior to cueing the speaker to ensure that the entire utterance is recorded onto tape. The experimenter and subject typically adopt a relatively rhythmic interchange by visual communication so that the utterances of a given list of sentences are recorded at a single sitting without verbal communication between the experimenter and subject after the first trial. After each list of sentences, the subject is encouraged to take a drink of water and relax for a short while. At midsession, the subject is asked to leave the chamber for about 5 min in order to minimize experimental fatigue.

The speaker is asked to refrain from placing contrastive or emphatic stress on any words or syllables in the utterance. If he does utilize such stress, or departs from his intended utterance in any way, he is asked to say "repeat" and then utter the sentence token again.

Acoustical analysis is then carried out on the first appropriate occurrence of each sentence. Except in rare cases of atypical stress pattern, the first token of each sentence is chosen for analysis. In the case of F_0 analysis, the procedure is quite complex, and a variety of available methods exist. We shall describe these below in an attempt to motivate our particular choice of F_0 analysis program.

Acoustical Measurement of F_0

In attempting to estimate the rate of any semiperiodic process, one must first decide over what time window the process should be measured in order to arrive at a good estimate. For example, if we say someone's heart is beating seventy times per minute, does that imply that his heartbeat is precisely 1.16 Hz (where Hz indicates cycles, or beats, per second)? If one accurately measures the duration of INDIVIDUAL heartbeats, this duration would vary somewhat from heartbeat to heartbeat. Hence, a difficulty arises in stating the overall "rate" of the process.

In the case of F_0, the duration of a complete opening and closing of the vocal cords varies from one individual cycle of open–close to the next. Over what total duration should we measure to obtain the rate of vibration of the vocal folds? One procedure could be to measure the duration of each individual open–close cycle (termed a "glottal cycle") and take the inverse, which gives a measure in units of rate.

Let us return to the example of heartbeat and measure the duration of ten consecutive heartbeats. Table 1.1 shows a hypothetical set of these durations. Assume that as soon as one heartbeat ends, the next begins; therefore, the total time for the heartbeats is simply the sum of the individual heartbeat durations. One way to describe the rate of heartbeat would be to simply state that the rate is 10 heartbeats/9.93 sec, or about 1.0075 heartbeats/sec, or 1.0075 Hz. Alternatively, one could take each individual heartbeat and simply take its inverse (e.g., for the fourth heartbeat, the rate is 1 heartbeat/1.01 seconds, or .99 Hz), also shown in Table 1.1.

Given this type of raw data (individual heartbeat durations), the task is to develop some way of expressing the rate of heartbeat, and, in particular, of expressing how the heartbeat rate changes as a function of time. Let us assume that, over the period of a single heartbeat, the rate is constant and equal simply to the inverse of the duration of the heartbeat. Thus, reading down column 3 of Table 1.1, the heartbeat is 1.031 Hz for .97 sec, then 1.020 Hz for the next .98 sec, then 1.000 Hz for the next 1.00 sec, and so forth. The advantage of this method of description is that it allows one to present the heartbeat rate as a plot of heartbeat frequency vs. time. The plot for the ten heartbeats is shown in Figure 1.9. When looking at this graph, it is important to realize that the vertical discontinuities do not represent sharp jumps in heartbeat rate at that point in time. Rather, Figure 1.9 simply represents the data from Table 1.1 in graphical form. Let us now return to F_0 and extend these concepts.

Shown in Figure 1.10 is a typical pressure amplitude vs. time plot of a portion of the voiced segment of a vowel. Note in particular the semiperiodic nature of the speech wave. This repeated portion of the waveform corresponds

Table 1.1. Data for heartbeat example

Heartbeat no.	Duration (sec)	Elapsed time (sec)	Inverse of duration (Hz)
1	.97	.97	1.031
2	.98	1.95	1.020
3	1.00	2.95	1.000
4	1.01	3.96	.990
5	1.00	4.96	1.000
6	.99	5.95	1.010
7	.97	6.92	1.031
8	.98	7.90	1.020
9	1.01	8.91	.990
10	1.02	9.93	.980

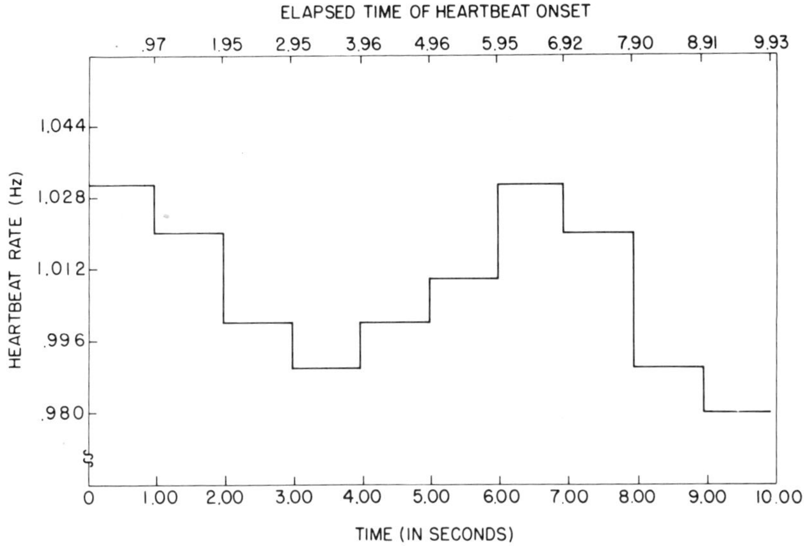

Figure 1.9. Plot of heartbeat frequency vs. time.

to the aforementioned glottal cycle. In order to measure F_0 for this speech sample, the first task is to measure the duration of each glottal cycle. Fortunately, there exists a measurable feature associated with the glottal cycle that is stable enough to be detectable in nearly all speech samples. This attribute is a correlate of the closure of the vocal cords, producing a long, sharply sloping "stroke" in the speechwave (Henke, 1974). We define "stroke" as a portion of the waveform between a local maximum and local minimum. (The slope of a stroke may be either positive or negative in polarity.) Once this long, sharply sloping stroke is located in each glottal cycle, the point in time where it crosses zero amplitude can be defined as the end of the previous glottal cycle and the beginning of the next one. The durations between successive points where this stroke crosses zero amplitude constitute the glottal cycle durations, which are then inverted and graphed as F_0 vs. time.

Dr. W. L. Henke has developed a program known as FPRD (FUNDAMENTAL PERIOD), which accomplishes the task outlined above. Specifically, FPRD searches for the sharply sloping strokes, and from them calculates the duration of individual glottal cycles. Another program then inverts each duration and

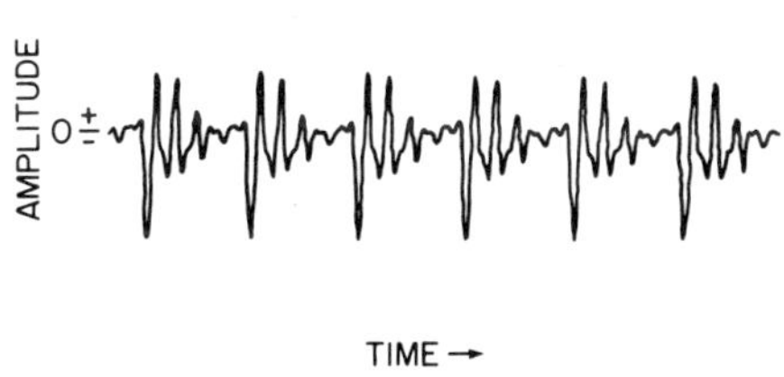

Figure 1.10. Plot of amplitude vs. time for a portion of a vowel.

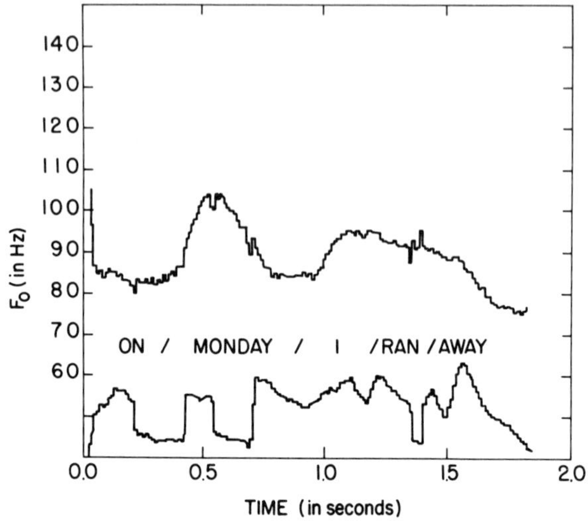

Figure 1.11. FPRD plot for the utterance *On Monday I ran away*. The associated amplitude pattern appears beneath the F_0 pattern on the display.

produces a plot of F_0 vs. time on a graphics display. A typical plot produced by FPRD appears in Figure 1.11. In addition to F_0, a measure of the relative energy (amplitude) associated with each glottal cycle is plotted directly below each F_0 value.

Let us now consider the accuracy of FPRD's duration measurements.[2] The input to the current implementation of FPRD is speech that has been analog low-pass filtered at 900 Hz and then digitized at 10 kHz. The 10 kHz sampling rate, combined with geometrical interpolation at the point where the strongest stroke crosses zero, provide a resolution in time of 1 μsec. Consider then a glottal cycle whose duration is 2000 μsec. The inverse of this duration is 500 Hz, certainly close to the upper limit of reasonable F_0 values, even for young children. A slightly shorter glottal cycle of 1999 μsec in duration inverts to give a value of 500.25 Hz. Even at this high F_0 level, the quantization is only one-fourth of one Hz. We can conclude that measuring the glottal cycle duration to the nearest microsecond is definitely accurate enough to neutralize any errors in measurement due to quantization.

While FPRD is well motivated by the considerations just outlined, this algorithm is by no means the only way of obtaining the F_0 contour for a given

[2]To illustrate the accuracy necessary in the calculation of glottal cycle duration, consider an example of a technique which measures such a duration to the nearest millisecond (one one-thousandth of a second). If one measures three typical glottal cycles with this technique and finds that their durations are 10, 9, and 10 msec, the corresponding measures of rate are 100 Hz, 111 Hz, and 100 Hz. Clearly, it will be unacceptable if the data are quantized such that the rate cannot change by less than 11 Hz. This problem worsens as the glottal cycle durations become smaller. A glottal cycle of 5 msec gives a corresponding rate of 200 Hz, and one of 4 msec gives 250 Hz. Hence, the measurement of duration must be considerably more accurate than to the nearest millisecond.

speech sample. Other algorithms are widely used and are motivated by particular applications (for a review and comparison of several algorithms for F_0 measurement, see Rabiner, Cheng, Rosenberg, and McGonegal, 1976). One common technique involves providing a measure of "estimate" of F_0 over a small portion or "window," of the input speech. This window is typically fixed for a given algorithm, and values of from 5–20 msec are often utilized for the window size. We have made use of one such program for the F_0 measurements of Study 3.1 (Gold and Rabiner, 1969), but all other new F_0 data presented in this book were produced by FPRD.

The FPRD program offers a number of attractive features for our use. First, it allows one to obtain values of F_0 at specified locations in a word. This feature cannot be accomplished by the programs which provide "estimates" of F_0 over a "window" of time, since it is usually not possible to specify the precise location in a word where the estimate of F_0 was calculated. Second, since the FPRD program indicates which strokes were chosen as the ending/beginning locations in the glottal cycles, the experimenter can determine the cause of occasional spurious F_0 values. Such errors usually occur in regions of speech accompanied by very low amplitude or in regions of fast F_0 transition, such as at the onset of voicing after a stop consonant. In our studies, we set an arbitrary criterion: to be considered as data, values of F_0 were required to be associated with an amplitude whose magnitude was at least 50% of the peak amplitude of the nearest vowel. If an F_0 value associated with a high amplitude was still suspected to be spurious because its value was more than 5 Hz above or below immediate neighboring values, then the experimenter reviewed the original speechwave to learn which strokes were (incorrectly) identified as the strongest strokes, and determined (with the aid of computer-controlled cursors) what the proper value(s) of F_0 should be for that portion of the speech signal. Finally, because FPRD operates in the time domain, it is computationally fast. In its current implementation on a now-obsolete ten-year-old PDP-9 computer, it requires only about 8 times real time to process input speech and obtain an F_0 contour. This rate is as fast or faster than most other time-domain programs. Typical frequency-domain algorithms require at least 30 times longer than FPRD to produce an F_0 contour (Rabiner et al., 1976).

Statistical Analysis

One of the distinguishing features of this particular study on F_0 involves the method of data reduction. In the past, most workers have tried to analyze F_0 contours by examining raw data points over the entire contour of an utterance. In this study, we measured preselected points of the contour which seemed on a priori grounds to be the most informative from a theoretical standpoint. In so doing, it was possible to analyze data for a few corresponding points over a larger range of sentences and speakers than has previously been attempted. Additionally, data reduction allowed us to apply statistical analyses to the results. This latter feature, although commonly employed in psychological studies, has not been applied in most previous studies of F_0. In the absence of any sort of data reduction and statistical treatment, the large amounts of data obtained in many earlier studies have been cumbersome and difficult to interpret. Indeed, some researchers (e.g., Crystal, 1969) have cited this drawback as a rationale for avoiding the process of data collection altogether. Yet, as Lieberman (1976) points out, this conclusion is tantamount to suggesting that astronomers should throw away their telescopes because so many stars are visible. It appears, rather, that one should make use of available specialized techniques in collecting data but then employ a theoretically motivated means of data reduction, even at a rudimentary stage of investigation, as is the current situation with F_0.

Summary

In this study, F_0 patterns are examined acoustically in utterances produced by groups of speakers in a sentence-reading task, designed to approximate spontaneous speech in as many respects as possible while at the same time permitting a high degree of experimental control over the sentence materials. With this type of test procedure, it is possible to test a number of hypotheses about the speaker's internal programming of F_0 attributes involving three major F_0 phenomena, including declination, fall–rise patterns, and the blocking of cross-word phonetic conditioning. These three phenomena were chosen for study because they seem to be influenced by the speaker's internal representation of syntactic structure.

The acoustical analysis of F_0 patterns is accomplished using a computer program that calculates the durations of individual glottal cycles and then inverts these durations to provide estimates of F_0. By using this program, it is possible to obtain highly accurate estimates of F_0 at particular locations in the utterance. The overall efficiency of the testing procedure and method of acoustical analysis permits us to examine a relatively broad range of issues with a number of different speakers. Whereas many previous studies of F_0 have examined a large corpus of data for only a few speakers, this study typically includes about 10 different speakers in each experiment, permitting statistical analyses of the data for the group of speakers as a whole. With this procedure, it is possible to determine which aspects of F_0 patterns are relatively speaker–independent.

Conclusion

The scientific study of F_0 has been introduced from the standpoints of linguistics, speech production, and acoustics. Our perspective combines these approaches in an attempt to study F_0 patterns in sentence contexts, with particular emphasis on the influencing role of syntactic structure.

Our general aim in this study is twofold. First, at least some of the salient aspects of F_0 patterns as they occur in sentence contexts must be adequately quantified for a relatively large range of sentence structures. Second, examining the characteristics of F_0 in sentence contexts should provide new information about how the speaker codes F_0 in the context of a variety of other attributes involved in speech production. For example, the study of F_0 should enable one to make inferences about how syntactic structures are represented by the speaker. In addition, the measurement of F_0 may provide new clues about how the speaker processes such structures in real time. The present experimental task is particularly well suited for this latter purpose, since the dependent variable constitutes an acoustic attribute that is included in the real-time activity of speech production itself.

2. Declination

The form and domain of F_0 declination are investigated here for a large variety of declarative sentence structures in English. The basic aim involves formulating a quantitative model of declination that is sufficiently abstract to capture the essence of this global attribute as it commonly appears in diverse sentence types. In so doing, we attempt to make inferences about the mental computations that influence the form of declination, and, in addition, provide a basis from which one can parcel out the effects of declination when attempting to study other, local F_0 attributes. Because declination appears so pervasively in speech, it seemed at the outset that a detailed analysis of this attribute was required as a foundation for studying other F_0 features such as fall–rise patterns and the operation of cross-word conditioning effects.

The dichotomy between global and local attributes of F_0 is central to any study of declination. By GLOBAL attributes, we shall refer to those characteristics of F_0 which can be best defined over a unit of speech that includes an entire clause or phrase. LOCAL attributes, on the other hand, represent those characteristics which can be best defined over a unit that includes no more than two adjacent words. In declarative speech, declination is the most salient global attribute, although other such characteristics are provided by stress effects defined over clause and phrase domains (e.g., Bresnan, 1971; Liberman and Prince, 1977). Local attributes include fall–rise patterns and effects of lexical stress, among others. For any given utterance, it is of course difficult to partition these global and local attributes. In this chapter, we shall nonetheless try to make such a partition, knowing full well that this attempt must be accompanied by at least a few arbitrary decisions about borderline instances of attributes that might be considered either global or local. In trying to characterize the distinction, however, a model of declination is formulated which seems to capture much of the essence of this attribute.

A fair amount of research has already attempted to characterize the F_0 declination pattern in declarative speech. Bolinger (1964), Cohen and t'Hart (1967), O'Shaughnessy (1976), and Maeda (1976), among others, document and describe F_0 declination. In the latter three studies, F_0 declination has been schematized as a single line with negative slope, drawn near or through the low values of F_0 occurring in an utterance. This reference line is typically termed the "baseline," and local patterns (e.g., fall–rise patterns) are defined with respect to this reference line (e.g., height above the baseline). However, given the F_0 contour of any particular utterance, it is very difficult to say with certainty through which low values the baseline should be drawn. Moreover, given any particular baseline, local patterns cannot be adequately defined with respect to it. In a very general sense, the F_0 contour shape does contain rises and falls, and descends from the beginning to the end of an utterance. However, for the purpose of trying to predict the salient characteristics of F_0 contours, the baseline model exhibits little predictive power. This drawback was quite evident in an extensive pilot study in which one of us (JMS) tried unsuccessfully to use the baseline model to schematize the F_0 contour of a speech corpus involving the reading of a well-known paragraph (The Rainbow Passage) by a group of eleven speakers.

Two other studies motivate the notion that two separate reference lines are required in a schematic description of F_0 contours. In one study (Breckenridge and Liberman, 1977), the authors state that a line which connects the peak values of F_0 in a sentence often declines faster than a line connecting the valleys. In another, O'Shaughnessy (1976) outlines a model for the F_0 contour which relies on two reference lines, one drawn through the successive peak values, the other through the F_0 valleys. Like Breckenridge and Liberman, he states that these two lines converge toward the end of the utterance and that both lines are monotonically decreasing.

These considerations, combined with our own preliminary observations of the general shape of F_0 contours, led to a preliminary experiment on F_0 declination. This experiment documented a difference in the slope of the line connecting the F_0 peaks and a line connecting the F_0 valleys in controlled sentences, as shown schematically in Figure 2.1.

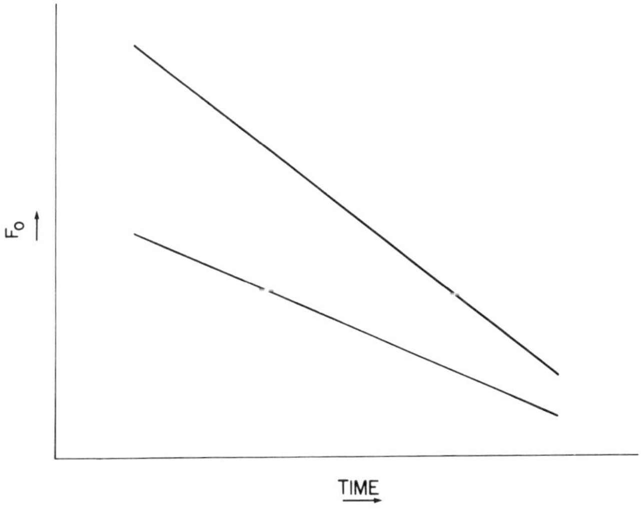

Figure 2.1. Schematic plot of F_0 vs. time showing the different slopes typically obtained for peaks and valleys in sentences.

Our approach was to place a carrier phrase at different positions within a sentence of constant overall length, hypothesizing that, if the topline and bottomline do approach one another, the distance between neighboring peak and valley values of F_0 should decrease when the carrier phrase is placed later in the sentence. The results for a group of nine speakers supported this notion.

Study 2.1. The F_0 Topline and the Topline Rule

Having obtained evidence that the topline and bottomline differ in their slopes, a detailed investigation of the topline was conducted. In selecting the topline for analysis, we considered a variety of criteria. This line was chosen because its associated F_0 peak values exhibited a variety of advantages over the bottomline, including relative ease of measurement, greater capability of being influenced by the speaker's coding of linguistic structures, and perceptual salience for the listener (see later).

Since the aim was to capture the essence of F_0 declination by some abstract mathematical formula, a significant issue involved normalization for individual differences among phonetic segments, speakers, and sentences. For example, the inherent F_0 of different vowels for male speakers varies over a range of nearly 20 Hz (Peterson and Barney, 1952; Lea, 1973). In order to derive abstract models for the F_0 topline, we constructed groups of sentences with similar syllabic and rhythmic structure and key words appearing at the same approximate locations in the different sentences of each group. The key words at a given sentence location across the different sentences contained different vowels. By calculating the average F_0 value at each key location for all sentences of a given group, it was thus possible to neutralize major differences in inherent vowel F_0. Since such an average typically represents 40 peak F_0 measurements (10 subjects speaking each of 4 key words), we expect a large standard deviation in these measurements. However, if the data base is large enough, the effects of individual speakers and inherent F_0 of different vowels is kept to a minimum. If the model has faithfully captured an essence of declination, it should be possible to predict the direction in which the model will fail in an individual utterance, based on prior knowledge of the key word, sentence structure, and speaker.

Experiment 2.1.1

The first experiment in this study involved two distinct groups of speakers and experimental sentences. Group 1 consisted of five male and five female subjects; Group 2, seven male and three female subjects; two of the male members of Group 1 also served as members of Group 2.

The sentences read by Group 1 are given below, with the key words underlined.[1] Note that the same three key words appear in the (a) and (b) versions of each sentence pair, with the (b) versions containing more material between the key words. By constructing the members of each sentence pair in this manner, we intended to provide a test of differences in peak F_0 in long vs. short sentences.

(1) a. *The deer could be seen from the car.*
b. *The deer by the canyon could be seen from the window of the car.*
(2) a. *The cat was asleep in the tree.*
b. *The cat that Sally owned was asleep on the large branch in the tree.*
(3) a. *The host could have packed all the plates.*
b. *The host at the party could have packed all the glasses with the plates.*
(4) a. *The fox has escaped from his cage.*
b. *The fox that Sidney caught has escaped sometime last night from his cage.*

The Group 2 sentences, with underlined key words, are as follows:

(5) a. *The fox was hiding from the dogs.*
b. *The fox in the hole was hiding from the hunters and their dogs.*
(6) a. *The weeds are growing in the field.*
b. *The weeds I will pull are growing near the flowers in the field.*
(7) a. *The teacher saw Joe.*
b. *The teacher in the parking lot saw a boy who looked like Joe.*
(8) a. *The students threw rocks.*
b. *The students in the parking lot threw a lot of sticks and rocks.*

The subjects were tested according to the procedure described in Chapter 1. The stressed syllable of each key word was analyzed using FPRD to calculate the peak value of F_0 and the time of occurrence of the peak relative to the beginning of the speech signal buffer. The peak values were labeled *P1*, *P2*, and *P3*, and the times of occurrence were labeled *T1*, *T2*, and *T3*. These times of occurrence were normalized, so that the first peak always occurred at *T1* = 1 sec. (Throughout the following discussion, uppercase "T" denotes a fixed location in time, while lowercase "t" denotes variable time.)

The aim of this first experiment was to determine the shape of the F_0 topline with a minimal amount of data. In addition, the matched pairs of short–long sentences would provide information about whether the shape of the topline differs with different sentence lengths.

Tables 2.1 (Group 1 data) and 2.2 (Group 2 data) contain the mean and standard deviation for each peak F_0 value and its time of occurrence in this experiment. These data are also shown graphically in Figures 2.2–2.5. Figure

[1]The results from Group 1 appeared originally in Sorensen and Cooper (1980).

Table 2.1. Means and standard deviations of values of peak F_0 (Hz) and associated time of occurrence (sec) for each of the key words in the group 1 sentences of experiment 2.1.1

Sentence			Measurement		
			P1,T1	*P2,T2*	*P3,T3*
(1)	a.	$\overline{F}_0$	201.	174.	160.
		SD	72.3	63.3	58.3
		$\bar{t}$	1.00	1.49	2.03
		SD	0.00	0.06	0.07
	b.	$\overline{F}_0$	213.	178.	154.
		SD	78.5	66.6	59.3
		$\bar{t}$	1.00	2.21	3.23
		SD	0.00	0.12	0.16
(2)	a.	$\overline{F}_0$	195.	171.	156.
		SD	65.0	55.1	59.9
		$\bar{t}$	1.00	1.54	2.04
		SD	0.00	0.06	0.08
	b.	$\overline{F}_0$	206.	182.	156.
		SD	75.5	63.8	61.8
		$\bar{t}$	1.00	2.30	3.75
		SD	0.00	0.12	0.17
(3)	a.	$\overline{F}_0$	205.	174.	154.
		SD	68.7	63.5	58.3
		$\bar{t}$	1.00	1.54	2.22
		SD	0.00	0.08	0.14
	b.	$\overline{F}_0$	220.	168.	155.
		SD	74.2	56.0	57.2
		$\bar{t}$	1.00	2.05	3.41
		SD	0.00	0.10	0.22
(4)	a.	$\overline{F}_0$	199.	178.	169.
		SD	70.7	65.5	65.7
		$\bar{t}$	1.00	1.52	2.19
		SD	0.00	0.06	0.10
	b.	$\overline{F}_0$	214.	178.	165.
		SD	79.5	65.9	64.9
		$\bar{t}$	1.00	2.33	3.94
		SD	0.00	0.13	0.15
Grand-means	a.	$\overline{F}_0$	200.	174.	159.
		SD	66.6	59.6	58.5
		$\bar{t}$	1.00	1.52	2.12
		SD	0.00	0.07	0.13
	b.	$\overline{F}_0$	213.	176.	157.
		SD	74.1	60.9	58.6
		$\bar{t}$	1.00	2.22	3.58
		SD	0.00	0.16	0.33

Table 2.2. Means and standard deviations of values of peak F_0 (Hz) and associated time of occurrence (sec) for each of the key words in the group 2 sentences of experiment 2.1.1

Sentence			Measurement		
			P1,T1	*P2,T2*	*P3,T3*
(5)	a.	$\bar{F}_0$	199.	182.	151.
		SD	65.9	55.8	49.0
		$\bar{t}$	1.00	1.42	2.06
		SD	0.00	0.09	0.11
	b.	$\bar{F}_0$	211.	183.	151.
		SD	72.1	58.4	46.1
		$\bar{t}$	1.00	1.94	3.25
		SD	0.00	0.08	0.17
(6)	a.	$\bar{F}_0$	203.	179.	170.
		SD	63.1	69.6	59.7
		$\bar{t}$	1.00	1.42	1.99
		SD	0.00	0.10	0.09
	b.	$\bar{F}_0$	228.	177.	166.
		SD	63.2	53.4	56.2
		$\bar{t}$	1.00	2.12	3.36
		SD	0.00	0.16	0.11
(7)	a.	$\bar{F}_0$	220.	181.	168.
		SD	76.0	65.5	55.9
		$\bar{t}$	1.00	1.37	1.70
		SD	0.00	0.19	0.18
	b.	$\bar{F}_0$	230.	166.	163.
		SD	74.3	54.3	56.6
		$\bar{t}$	1.00	2.16	3.33
		SD	0.00	0.15	0.23
(8)	a.	$\bar{F}_0$	211.	178.	157.
		SD	66.4	54.9	49.8
		$\bar{t}$	1.00	1.42	1.66
		SD	0.00	0.07	0.08
	b.	$\bar{F}_0$	227.	182.	150.
		SD	71.4	65.0	51.3
		$\bar{t}$	1.00	2.19	3.20
		SD	0.00	0.18	0.26
Grand-means	a.	$\bar{F}_0$	208.	180.	161.
		SD	65.9	59.4	52.3
		$\bar{t}$	1.00	1.41	1.85
		SD	0.00	0.12	0.21
	b.	$\bar{F}_0$	224.	177.	157.
		SD	68.0	56.1	51.2
		$\bar{t}$	1.00	2.10	3.28
		SD	0.00	0.18	0.20

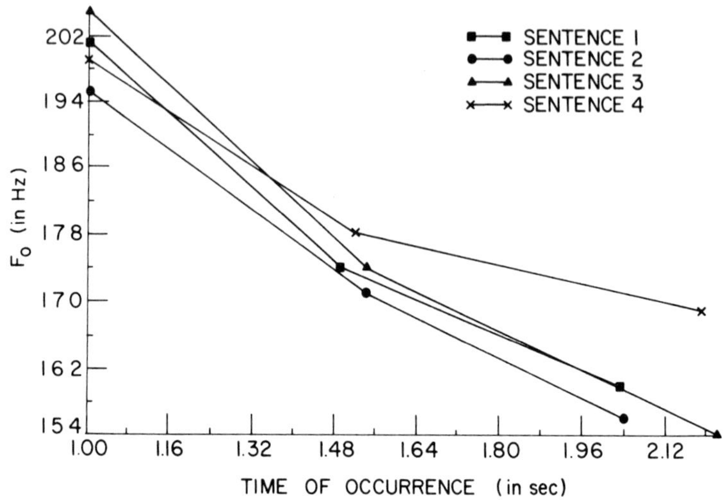

Figure 2.2. Plot of mean F_0 vs. time for each of the measured peak values in Sentences (1a)–(4a).

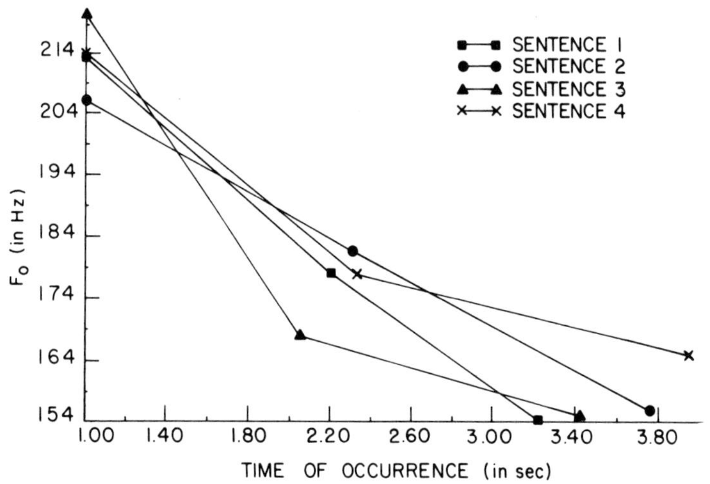

Figure 2.3. Plot of mean F_0 vs. time for each of the measured peak values in Sentences (1b)–(4b).

2.6 shows the mean peak F_0 values and the mean times of occurrence for all four types of sentences.

After several attempts to characterize these curves, the following approximation emerged:

> For any one curve, locate the point halfway between *T1* and *T3* on the time axis, and one-third of the distance up from *P3* toward *P1* on the frequency axis (since we shall need to refer to this point on the graph fairly often, we denote it with an asterisk, "*".

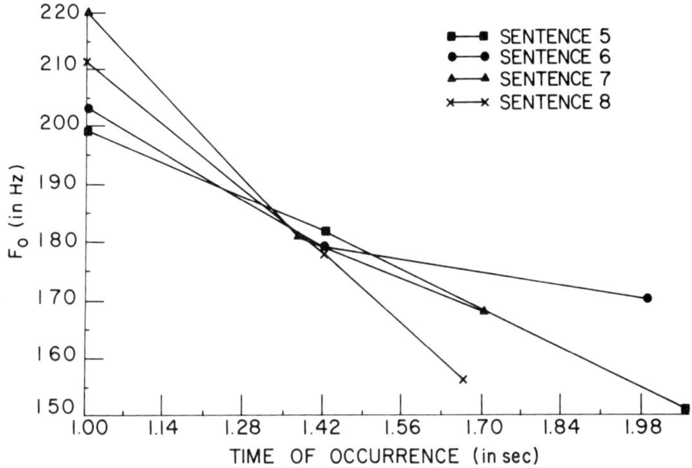

Figure 2.4. Plot of mean F_0 vs. time for each of the measured peak values in Sentences (5a)–(8a).

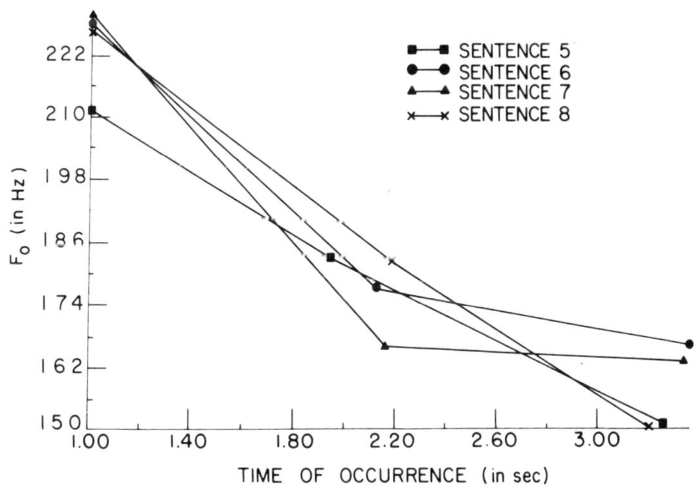

Figure 2.5. Plot of mean F_0 vs. time for each of the measured peak values in Sentences (5b)–(8b).

The line segments from (*T1*,*P1*) and (*T3*,*P3*) to "*" constitute an approximation to the shape of the topline, based on the data from this experiment. Algebraically, the two line segments describe a relationship between time, the independent variable, and F_0, the dependent variable, as follows. For the segment connecting (*T1*,*P1*) to "*",

$$F_0 = P1 + \frac{4}{3}\left(\frac{P1 - P3}{T1 - T3}\right)(t - T1). \tag{2.1}$$

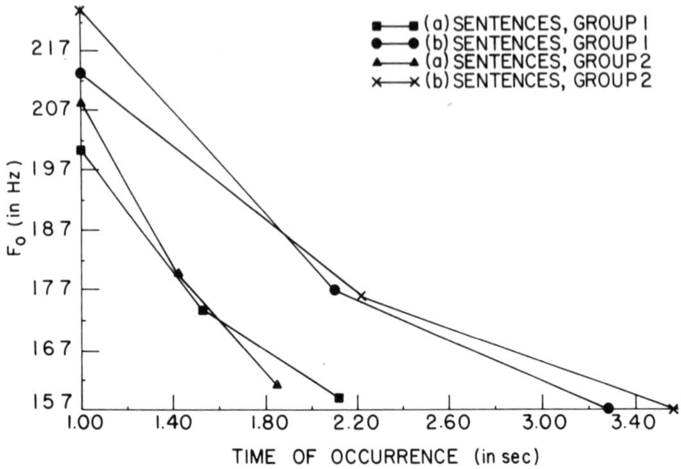

Figure 2.6. Plot of mean F_0 vs. time for each of the measured peak values pooled across sentences for the (a) and (b) sentences of Groups 1 and 2.

For the segment connecting "*" to ($T3$, $P3$),

$$F_0 = P3 + \frac{2}{3}\left(\frac{P1 - P3}{T1 - T3}\right)(t - T3). \tag{2.2}$$

For the general case of a sentence with n measured F_0 peaks, the equations become, from ($T1,P1$) to "*",

$$F_0 = P1 + \frac{4}{3}\left(\frac{P1 - Pn}{T1 - Tn}\right)(t - T1). \tag{2.3}$$

From "*" to (Tn,Pn), we have

$$F_0 = Pn + \frac{2}{3}\left(\frac{P1 - Pn}{T1 - Tn}\right)(t - Tn). \tag{2.4}$$

The F_0 output from these equations is shown graphically in Figure 2.7. These equations imply that, given the location and F_0 value of the first ($T1,P1$) and final (Tn,Pn) F_0 peaks, one can predict the value of peak F_0 at any other location t. With the present data, one can predict a value of $P2$, occurring at time t, given ($T1,P1$) and ($T3,P3$) of an individual sentence.

An additional analysis was carried out to assess differences in the magnitude of the actual F_0 values, depending on the overall sentence length. Table 2.3 contains the results of a series of two-tailed t tests for matched pairs, comparing the values of peak F_0 of matched key words in the long vs. short versions of all eight sentence pairs in this experiment. The t tests indicate that the values of $P1$ in the long versions were significantly higher than the values of $P1$ in the short versions in seven out of eight sentence pairs. The only nonsignificant pair, Sentence Pair (2), showed a nearly significant trend in the same direction ($t = 2.07$, $.07 > p > .05$, $df = 9$). The magnitude of this difference, averaged across all eight sentence pairs, was 14.5 Hz.

In sharp contrast, the t tests on the matched values of later peaks, $P2$ and

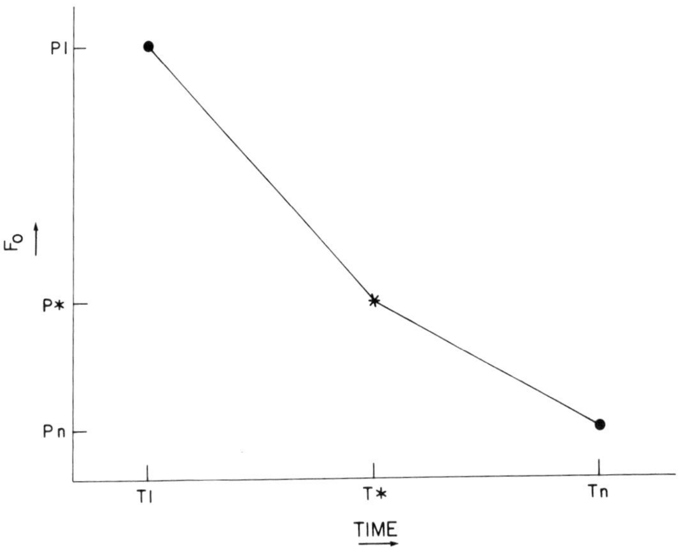

Figure 2.7. Plot of F_0 vs. time output for Equations 2.1–2.4 (Topline Model 1).

P3, showed little or no difference. While Sentence Pairs (2) and (7) yielded significant differences in the values for *P2*, the differences were in the opposite direction from one another. The matched values of *P3* showed no significant differences. On average, the values for *P2* and *P3* in the long sentence versions were .38 Hz and 3.13 Hz lower, respectively, than in the short sentence versions.

The slope of the F_0 topline was much steeper in short vs. long sentences as a consequence of the differences in sentence durations and values of *P1*. The long sentences were about 100% longer than the short sentences, whereas *P1*

Table 2.3. t Values for peak F_0 comparisons in long vs. short sentences in experiment 2.1.1

Sentence pair	*P1*	*P2*	*P3*
(1)	2.67[a]	0.99	−2.16
(2)	2.07	2.62[a]	−0.04
(3)	3.78[c]	−1.11	0.21
(4)	2.94[b]	−0.10	−1.05
(5)	2.98[b]	0.23	−0.03
(6)	5.79[c]	−0.37	−0.69
(7)	2.98[b]	−2.28[a]	−1.79
(8)	2.78[a]	0.41	−1.74

[a] $p < .05$.
[b] $p < .02$.
[c] $p < .01$.

in the long sentences averaged only about 6% higher. Consequently, the declination of F_0 peaks is produced neither at a constant rate nor by a constant overall amount for sentences of different lengths. The declination amount is more nearly constant, however, than is the declination rate.

The higher values for *P1* in longer sentences suggest that the speaker employs a look-ahead mechanism which computes the approximate overall upcoming utterance length (see also Experiment 2.2.1). The finding that F_0 tends to start higher in longer utterances has also been noted by McAllister (1971) and O'Shaughnessy (1976). On the basis of these data alone, however, it is not possible to determine whether this planning effect takes into account the entire length of the upcoming sentence, or, for example, only the length of the first major phrase. The *P1* effect also raises other interesting questions about the precise characterization of the speaker's look-ahead routine, but these questions must await empirical testing in further work.

While the finding of higher values for *P1* in longer utterances appears consistently in our data, this effect is not found when speakers are asked to utter a prespecified string of UNRELATED words under time pressure (Sternberg, Wright, Knoll, and Monsell, 1980). It thus appears that the effect for *P1* might be restricted to meaningful fluent speech spoken at relatively normal rates, even though the more general attribute of declination is observed with both fluent speech and unrelated words (Sternberg et al., 1980). In examining F_0 declination, Sternberg et al. found that the rate of declination decreased as a function of utterance length, in general accord with our results. Thus, unlike the *P1* effect, the effect of utterance length on declination rate does appear to reflect a general property of motor programming regardless of whether or not the utterance is a meaningful unit.

Experiment 2.1.2

In this experiment, we sought to observe whether other types of sentences exhibited an F_0 topline similar to the one modeled in Experiment 2.1.1. The test sentences contained four measurable peaks instead of three, in order to examine the characteristics of the topline in more detail. In analyzing this experiment, two refinements of the original topline model were formulated, and we assessed how well any one model described the data of individual sentences.

Twenty subjects participated in this experiment, consisting of 11 male and 9 female subjects, none of whom served in Experiment 2.1.1.[2] The sentence materials for this experiment are shown below, with the key words underlined:

(9) *The <u>book</u> on the <u>table</u> was a <u>gift</u> from my <u>mother</u>.*
(10) *The <u>man</u> in the <u>truck</u> sent <u>roses</u> to his <u>niece</u>.*
(11) *The <u>students</u> on the <u>bus</u> threw <u>gum</u> at the <u>teacher</u>.*
(12) *The <u>ducks</u> in the <u>park</u> will eat <u>popcorn</u> and <u>peanuts</u>.*

[2]The results for the first 10 subjects appeared originally in Sorensen and Cooper (1980).

Table 2.4. Means and standard deviations of values of peak F_0 (Hz) and associated time of occurrence (sec) for each of the key words in the sentences of experiment 2.1.2

Sentence		Measurement			
		P1,T1	*P2,T2*	*P3,T3*	*P4,T4*
(9)	$\bar{F}_0$	187.	173.	159.	142.
	SD	49.3	44.7	42.5	40.2
	$\bar{t}$	1.00	1.45	2.04	2.64
	SD	0.00	0.05	0.11	0.16
(10)	$\bar{F}_0$	179.	165.	157.	148.
	SD	50.0	44.1	41.5	40.6
	$\bar{t}$	1.00	1.40	1.97	2.63
	SD	0.00	0.05	0.10	0.12
(11)	$\bar{F}_0$	200.	158.	155.	152.
	SD	59.9	41.6	41.3	42.0
	$\bar{t}$	1.00	1.50	2.05	2.50
	SD	0.00	0.06	0.09	0.11
(12)	$\bar{F}_0$	184.	171.	161.	155.
	SD	54.6	45.6	42.9	41.7
	$\bar{t}$	1.00	1.50	2.11	2.73
	SD	0.00	0.07	0.11	0.16
Grand-means	$\bar{F}_0$	187.	167.	158.	149.
	SD	53.2	43.6	41.3	40.6
	$\bar{t}$	1.00	1.46	2.04	2.62
	SD	0.00	0.07	0.11	0.16

Three matched filler sentences accompanied the test sentences on the list presented to each subject. As in previous experiments, the first appropriate token of each utterance for all subjects was analyzed using the FPRD program. The peak F_0 and time of occurrence were recorded for each key word. Table 2.4 shows the mean values for F_0 and normalized time of occurrence for each key word in each sentence, as well as the grand-mean and standard deviations across all sentences. The mean data for all four sentences is shown in Figure 2.8.

The influence of different phonetic structures on F_0 can readily be seen in this figure. For example, the value of *P1* in Sentence (11) was about 10 Hz higher than the values of *P1* for the other three sentences. The higher value of *P1* in (11) can be attributed to the fact that the vowel /u/ in the stressed syllable of the key word *students* exhibits a higher inherent F_0 value than any of the other vowels in *P1* words of other sentences (Peterson and Barney, 1952, Mohr, 1971). Moreover, this vowel is immediately preceded by a voiceless stop in (11), unlike the vowels in the other three sentences, and the influence of this voiceless stop would also produce higher values for the vowel (e.g., House and

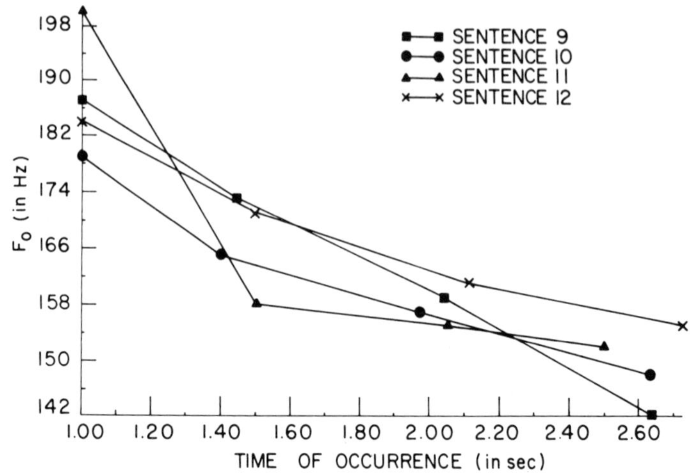

Figure 2.8. Plot of mean F_0 vs. time for each of the measured peak values in Sentences (9)–(12).

Fairbanks, 1953; Lea, 1973), as reviewed in Chapter 1. The influence of phonetic structure can also partially account for the fact that the *P2* word *bus* in Sentence (11) exhibited the lowest F_0 values of the four sentences. Unlike the other *P2* words, *bus* begins with a voiced consonant, yielding relatively low F_0 values during the following vowel.

Figure 2.9 shows a plot of the grand-mean data for this experiment, and the model for the F_0 topline, based on Equations 2.3 and 2.4, with $n = 4$. The point (*T3,P3*) lies almost precisely on the line given by the model. From "*" to (*T4,P4*), this model adequately describes the actual F_0 topline. However, this is not the case from (*T1,P1*) to "*". The (*T2,P2*) point is well below the

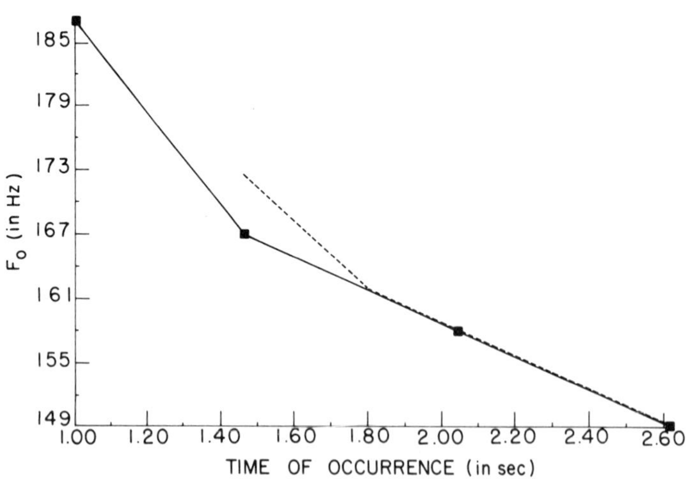

Figure 2.9. Predicted (dotted line) and obtained (solid line) values of mean F_0 vs. time for the grand-mean data of Experiment 2.1.2, using Topline Model 1 to derive the predicted values.

value predicted by the model. The mismatch is not surprising, since the previous experiment did not provide any data about the shape of the topline in the regions close to *T1*. These new data prompted the following refinement in the model:

> Move "*" so that it is located halfway between *P1* and *Pn* on the frequency axis and one fourth of the distance from *T1* to *Tn* on the time axis.

The two segments connecting "*" to (*T1*,*P1*) and (*Tn*,*Pn*) constitute the revised model to the topline. The F_0 output for this second topline is shown in Figure 2.10. The segment from (*T1*,*P1*) to "*" is described by the following equation:

$$F_0 = P1 + 2\left(\frac{P1 - Pn}{T1 - Tn}\right)(t - T1). \tag{2.5}$$

The segment connecting "*" to (*Tn*, *Pn*) is mathematically unchanged from Equation 2.4; however, we repeat it here for convenience:

$$F_0 = Pn + \frac{2}{3}\left(\frac{P1 - Pn}{T1 - Tn}\right)(t - Tn). \tag{2.6}$$

Both the first model for the topline and this revised model are comprised of two line segments. However, we see also in Figure 2.9 that the points from (*T2*,*P2*) to (*Tn*,*Pn*) lie essentially along a straight line, while the first peak (*T1*,*P1*) lies well above this line. It is not obvious why the mean value *P1* lies above the line given by the other three peaks. Quite possibly, the high value of *P1* represents some type of start-up effect at the beginning of a new utterance (see also Experiment 2.3.1). This same effect has been noted by others (e.g.,

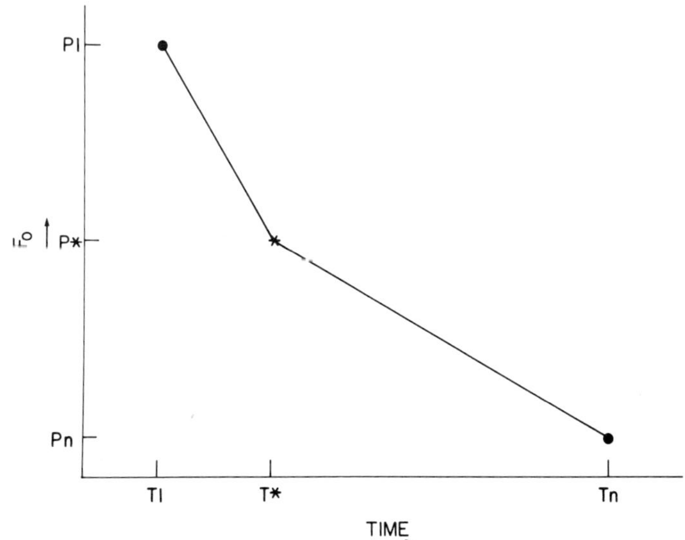

Figure 2.10. Plot of F_0 vs. time output for Topline Model 2.

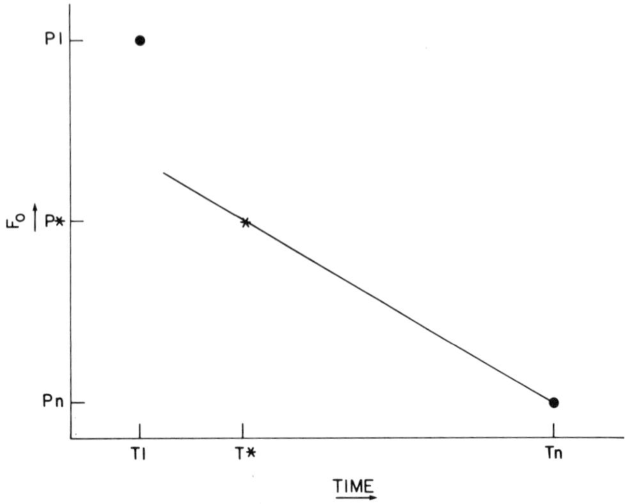

Figure 2.11. Plot of F_0 vs. time output for Topline Model 3. This version will be adopted throughout the remainder of this chapter as the model that best represents the F_0 topline.

Olive, 1975; Maeda, 1976). To simplify the topline model, it was decided to allow ($T1,P1$) to appear above the line and extend the second line segment back to the location in time of $T2$. This single-line model is represented in Figure 2.11. In this model, the peak F_0 at any intermediate location in an utterance is predicted by Equation 2.6.

Finally, each of the three topline models of individual sentences was examined to uncover how accurately it predicted the data of individual sentences. These tests involved comparing the actual values of F_0 vs. the values predicted by each of the three models. From each comparison, a percent of error calculation can be made, based on how well the model predicted a particular peak F_0 value.

This procedure will be illustrated by calculating the percent of error terms for the data of Sentence (9), spoken by Subject 1, a female. The raw data and the error calculations are shown in Table 2.5. Note that the data include four peak F_0 values and four associated times of occurrence, corresponding to the four key words in Sentence (9). The topline model to be used in this example is the third model, shown graphically in Figure 2.12. Recall that, in this model, the point ($T1,P1$) lies well above the line, and the model is described by Equation (2.6) at all intermediate locations along the utterance. Line (1) in Table 2.5 shows Equation (2.6), in which we have substituted the actual values of $T1,Pl$, Tn, and Pn. Line (2) shows a simplification of (1); the relationship between peak F_0 and the independent variable time (t) is more clearly evident. Note also that the equation on Line (2) describing F_0 as a function of time is applicable ONLY to this individual utterance. From the equation of Line (2), one can now predict a peak F_0 value at a specified location. Line (3) shows that, at $T2$ = 1.42 sec, the model predicts the value of peak F_0, $P2$, to be 233.94 Hz. The raw data at the top of Table 2.5 indicate that, at this location,

Table 2.5. Example error term calculation

Key word	Peak F_0	Time of occurrence (sec)
1	270	1.00
2	239	1.42
3	210	1.97
4	200	2.54

From Equation (2.6), we have the following:

$$F_0 = 200 + \frac{2}{3}\left(\frac{270 - 200}{1.00 - 2.54}\right)(t - 2.54). \tag{1}$$

$$F_0 = 276.97 - 30.30(t). \tag{2}$$

We can now predict peak F_0 at time *T2* (i.e., predict the value of *P2*):

$$\begin{aligned} F_0 &= 276.97 - (30.30)(1.42); \\ &= 233.94 \text{ Hz.} \end{aligned} \tag{3}$$

The percent of error in the prediction is defined as:

$$\text{Error} = \frac{F_0 \text{ obtained} - F_0 \text{ predicted}}{F_0 \text{ predicted}}.$$

Thus, the error in our estimate of F_0 at time *T2* is:

$$\frac{239 - 233.94}{233.94} = .0216, \text{ or } 2.16\%. \tag{4}$$

Similarly, predicting peak F_0 at time *T3* (i.e., predicting *P3*);

$$\begin{aligned} F_0 &= 267.97 - (30.30)(1.97); \\ &= 217.27 \text{ Hz.} \end{aligned} \tag{5}$$

$$\begin{aligned} \text{Error} &= \frac{210 - 217.27}{217.27} \\ &= -.0335, \text{ or } -3.35\%. \end{aligned} \tag{6}$$

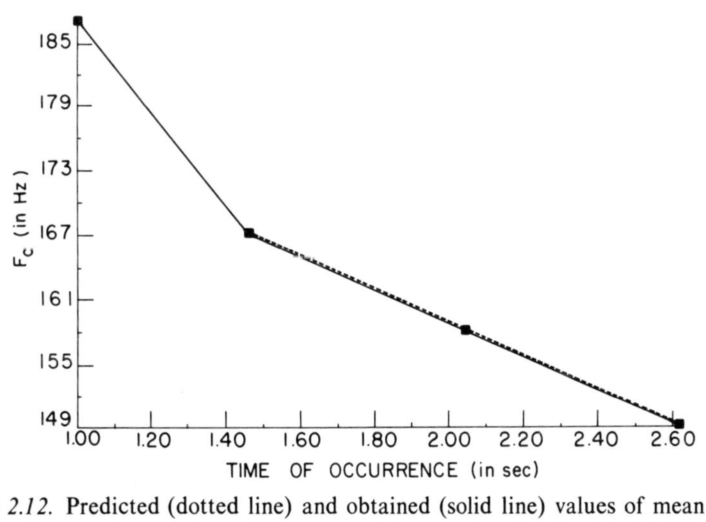

Figure 2.12. Predicted (dotted line) and obtained (solid line) values of mean F_0 vs. time for the grand mean data of Experiment 2.1.2, using Topline Model 3 to derive the predicted values.

the measured value of *P2* was 210 Hz. The predicted value is compared with the actual value in Line (4), the percent of error being 2.16%. Lines (5)–(6) repeat the procedure of Lines (3)–(4), predicting the value of *P3* at the location *T3* and calculating the percent error term, in this case, −3.35%. Positive error terms indicate that the actual F_0 values lie above the line produced by the model, whereas negative error terms signify the actual peak F_0 values lie below this line.

In Table 2.6, we present the complete set of percent error terms for Sentences (9)–(12) for Subject 1, based again on the single-line segment version of the topline model. Of particular importance is the method of calculating the mean percent of error terms. This calculation was performed while RESPECTING the sign of the individual error terms, such that the mean percent of error equals the arithmetic sum of the terms divided by the number of terms. In the case of the error terms for *P2* in Table 2.6, the arithmetic sum of (+2.16 − 5.72 − 11.38 − 12.82) is −27.76, and −27.76/4 = −6.94. To further illustrate the implication of this method of calculation, consider the percent error terms of *P2* for Subject 19 for Sentences (9)–(12). The individual percent of error terms are +0.27, +2.99, −5.20 and +2.61%. The arithmetic sum of these four terms is +.67%, and the mean is therefore 0.17%. The mean percent of error in this example is considerably smaller in magnitude than ANY of the individual error terms, resulting from the fact that positive and negative error terms, when equal in magnitude, cancel each other out.

The respecting of sign is adopted in the calculation of the mean error term because the mean error is intended to reflect the degree to which the topline model captures the essence of declination. In any individual utterance, as noted earlier, the values of the F_0 peaks may be influenced by factors other than declination to such an extent that the individual error terms for the utterance are quite large. Since the topline model is not intended to capture the influence of factors other than declination, the model is poor in predicting the peak values in such utterances. Because the topline model is designed to represent declination, a fair test of its predictive power can be provided by measuring the mean percent error across a variety of utterances. In such a test, the positive

Table 2.6. Error terms for subject 1, experiment 2.1.2

	% error in prediction of peak	
Sentence	*P2*	*P3*
(9)	2.16	−3.35
(10)	−5.72	−4.22
(11)	−11.38	−5.64
(12)	−12.82	−9.08
Mean	−6.94	−5.57
SD	6.80	2.52

and negative errors due to other factors should cancel out for the most part and leave declination as the principal distillate.

Table 2.7 contains the complete set of error terms for all twenty subjects in this experiment. Two error terms are given for each utterance, these terms corresponding to the error in the prediction of the values of *P2* and *P3*, respectively, using the single-line topline model. The mean and standard deviations for the error terms across the four sentences are shown for each subject. Table 2.7 also gives the mean and standard deviation for the error terms for each sentence across all twenty subjects, and finally, the overall mean and standard deviation for all predictions of *P2* and *P3*.

As Table 2.7 shows, the single-line topline model predicts the mean peak F_0 values for male and female speakers with about the same degree of accuracy, despite the fact that the absolute values of these peaks were approximately 1.8 times higher for female speakers. For the eleven males, the error terms for *P2* and *P3* averaged 1.52% and 0.86%, respectively. For the nine females, these error terms averaged −1.58% and −0.70%. The capability of the single-line topline model to predict male and female F_0 peaks with approximately the same high degree of accuracy was observed not only here but throughout the experiments contained in this chapter.

We come now to the task of determining which of the three models most accurately represents the topline on the basis of the data gathered so far. To begin, we can eliminate the first model, consisting of two segments and shown earlier in Figure 2.7. As indicated for the data for Experiment 2.1.2, this model predicts peak F_0 values that are much too high for locations between (*T1*,*P1*) and the midpoint of the utterance. We are thus left with two plausible models. Model 2, shown in Figure 2.10, contains two line segments, and Model 3, shown in Figure 2.11, contains only one line segment. The segment in Model 2 from "*" to (*Tn*,*Pn*) is colinear with Model 3. However, from (*T1*,*P1*) to "*", Model 2 predicts higher values of peak F_0 than Model 3.

Two basic characteristics appear to favor Model 3. The first involves simplicity, there being little advantage to using two segments to model the F_0 topline if it can be accomplished almost as precisely by one segment. The second involves the strongly linear relationship among (*T2*,*P2*), (*T3*,*P3*) and (*T4*,*P4*) (see Figure 2.9). A two-segment version of the topline model (Model 2) tends to obscure this important characteristic of the topline. When only one line segment is used for a topline model (Model 3), the basic features of the F_0 topline are readily apparent: the points from (*T2*,*P2*) to (*Tn*,*Pn*) are related in linear fashion, and (*T1*,*P1*) does not lie on this same line. One may add here that we could have just as easily used (*T2*,*P2*) and (*Tn*,*Pn*) to predict (*T1*,*P1*) and any remaining peaks. This procedure was not chosen because we wanted the defining values in the topline rule equation (Equation 2.6) to represent the temporal endpoints (and, in most cases, the highest and lowest peak F_0) in the sentence.

But how does the accuracy of each model, based on the mean error terms, aid in determining which model is better? Recall Table 2.7, where we presented the complete set of error terms for all twenty subjects of Experiment 2.1.2. The grand-mean percent of error terms, across all subjects and sentences, were .13% and .16% for *P2* and *P3*, respectively, based on Model 3. The grand-

Table 2.7. Individual, mean, grand, and grand-grand mean percent of error terms for the two predicted vs. obtained peak F_0 values in the sentences of experiment 2.1.2

	% error in prediction of peak			% error in prediction of peak	
	P2	*P3*		*P2*	*P3*
Subject 1[a]			Subject 6[b]		
Sentence			Sentence		
(9)	2.16	−3.35	(9)	12.88	3.30
(10)	−5.72	−4.22	(10)	10.35	−1.53
(11)	−11.38	−5.64	(11)	−2.67	−0.64
(12)	−12.82	−9.08	(12)	4.56	6.65
Mean	−6.94	−5.57	Mean	6.28	1.95
SD	6.80	2.52	SD	6.91	3.77
Subject 2[b]			Subject 7[b]		
Sentence			Sentence		
(9)	11.39	1.02	(9)	15.27	14.29
(10)	15.74	5.94	(10)	6.91	6.84
(11)	−5.46	−3.39	(11)	0.69	−3.14
(12)	7.33	3.67	(12)	−4.14	0.69
Mean	7.25	1.81	Mean	4.68	4.67
SD	9.14	4.01	SD	8.38	7.62
Subject 3[b]			Subject 8[b]		
Sentence			Sentence		
(9)	2.80	3.65	(9)	16.34	4.39
(10)	5.95	7.88	(10)	7.36	4.92
(11)	−9.01	−9.69	(11)	−5.59	−2.45
(12)	−2.57	−15.24	(12)	7.11	4.10
Mean	−0.71	−3.35	Mean	6.30	2.74
SD	6.56	10.90	SD	9.02	3.47
Subject 4[b]			Subject 9[a]		
Sentence			Sentence		
(9)	4.81	2.48	(9)	−3.12	4.77
(10)	3.57	3.10	(10)	−0.47	−0.70
(11)	−2.86	−4.14	(11)	−6.57	3.07
(12)	2.70	−1.65	(12)	−8.98	−5.80
Mean	2.06	−0.05	Mean	−4.78	0.33
SD	3.39	3.45	SD	3.75	4.69
Subject 5[b]			Subject 10[a]		
Sentence			Sentence		
(9)	−7.08	1.18	(9)	4.85	1.78
(10)	−8.73	−3.56	(10)	2.12	−3.50
(11)	−10.60	−0.01	(11)	−12.45	−9.50
(12)	−9.61	−2.03	(12)	0.24	−4.72
Mean	−9.00	−1.10	Mean	−1.31	−3.99
SD	1.49	2.10	SD	7.66	4.63

Table 2.7. continued

	% error in prediction of peak			% error in prediction of peak	
	P2	*P3*		*P2*	*P3*
Subject 11[a]			Subject 16[a]		
Sentence			Sentence		
(9)	10.99	0.99	(9)	−0.74	3.91
(10)	8.20	−2.20	(10)	−2.24	4.81
(11)	−6.98	−7.36	(11)	−13.68	−4.46
(12)	7.50	0.54	(12)	2.20	8.41
Mean	4.93	−2.01	Mean	−3.61	3.17
SD	8.08	3.84	SD	6.96	5.44
Subject 12[a]			Subject 17[b]		
Sentence			Sentence		
(9)	7.35	7.12	(9)	17.40	7.63
(10)	−5.45	2.24	(10)	−2.02	2.70
(11)	−10.48	−3.93	(11)	−16.33	−0.34
(12)	3.40	3.22	(12)	19.37	4.46
Mean	−1.29	2.16	Mean	4.61	3.61
SD	8.14	4.58	SD	16.97	3.33
Subject 13[a]			Subject 18[b]		
Sentence			Sentence		
(9)	6.43	2.48	(9)	4.92	1.98
(10)	3.78	0.80	(10)	−13.38	−9.01
(11)	−7.66	−2.72	(11)	−13.11	−9.85
(12)	1.42	−0.22	(12)	−4.23	5.60
Mean	0.99	0.09	Mean	−6.45	−2.82
SD	6.12	2.18	SD	8.69	7.78
Subject 14[a]			Subject 19[b]		
Sentence			Sentence		
(9)	5.57	10.83	(9)	0.27	0.86
(10)	−4.02	−2.78	(10)	2.99	1.92
(11)	−11.87	−3.52	(11)	−5.20	−0.57
(12)	10.29	4.55	(12)	2.61	−2.97
Mean	−0.01	2.27	Mean	0.17	−0.19
SD	9.90	6.77	SD	3.78	2.11
Subject 15[a]			Subject 20[b]		
Sentence			Sentence		
(9)	3.53	3.24	(9)	4.93	4.79
(10)	0.69	−2.14	(10)	0.49	2.74
(11)	−13.16	−9.03	(11)	−5.31	−0.28
(12)	0.02	−3.12	(12)	6.18	1.44
Mean	−2.23	−2.76	Mean	1.57	2.17
SD	7.44	5.03	SD	5.20	2.14

Table 2.7. continued

		% error in prediction of peak	
		P2	*P3*
Grand-means and SDs			
(9)	$\bar{E}$	6.05	3.87
	SD	6.47	3.85
(10)	$\bar{E}$	1.31	0.71
	SD	6.92	4.35
(11)	$\bar{E}$	−8.48	−3.88
	SD	4.42	3.68
(12)	$\bar{E}$	1.63	−0.08
	SD	7.47	5.74
Overall means and SDs			
	$\bar{E}$	0.13	0.16
	SD	8.26	5.20

[a]female
[b]male

mean percent error terms based on Model 2 were .04% and .16% for *P2* and *P3*. As pointed out above, Model 2 predicts higher values of F_0 during the first one-fourth of an utterance than Model 3. Since all the values of *T3* occurred after "*" in the individual utterances, the predicted values of *P3* were the same for both models in every case. The grand-mean percent error term for *P3* for both models was noted above as .16%. However, the two models produced a differing grand-mean percent error for *P2*. Approximately one-fourth of the values of *T2* occurred prior to "*" in the individual utterances. In such instances, the error terms differed for the two models. For Model 2, the grand-mean percent error was .04%, which implies that the "average" actual value was slightly ABOVE the topline predicted by this model. Since Model 3 predicts a lower topline in the region to the left of "*", it follows that the mean percent error for Model 3 would be slightly higher, in fact at .13%. From this set of data alone, the analysis of the grand-mean percent error terms would favor Model 2 as the more accurate topline model. Considering the data for a number of other sentence groupings studied elsewhere in this chapter where values of *T2* occasionally occurred prior to "*", Model 2 showed a smaller overall percent of error in seven cases, and Model 3 was closer in the remaining three cases. However, over these ten cases, the Model 2 mean percent error terms for *P2* averaged only .075% smaller than those given by Model 3. Given this slight difference and the attractiveness of Model 3 from the standpoints of simplicity and clarity, we have opted to utilize Model 3 as our approximation to the F_0 topline and have named it the TOPLINE RULE.

Table 2.8. Mean and grand-mean percent of error terms of the one predicted vs. obtained peak F_0 value in the group 1 (a) sentences of experiment 2.1.1

Sentence			% error in prediction of *P2*
(1)	a.	$\overline{E}$	−0.09
		SD	7.49
(2)	a.	$\overline{E}$	2.34
		SD	5.44
(3)	a.	$\overline{E}$	0.92
		SD	6.99
(4)	a.	$\overline{E}$	−0.99
		SD	6.49
Grand-mean	a.	$\overline{E}$	0.55
		SD	6.51

Figure 2.12 illustrates the grand-mean data for Experiment 2.1.2, along with the adopted Topline Rule calculated from the grand-mean values of (*T1*,*P1*) and (*T4*,*P4*). Excellent Topline Rule approximations to the grand-mean values of *P2* and *P3* appear in Figure 2.12, consistent with the grand-mean percent of error terms at the bottom of Table 2.7.

Before moving on to the next study, we present the mean percent error terms and grand-mean data with the associated Topline Rule for the four different sentence groupings of Experiment 2.1.1. The tables and figures are organized as follows: Grand-mean error terms: Experiment 2.1.1, Group 1 (a) sentences in Table 2.8, (b) sentences in Table 2.9, Group 2 (a) sentences in Table 2.10,

Table 2.9. Mean and grand-mean percent of error terms of the one predicted vs. obtained peak F_0 value in the group 1 (b) sentences of experiment 2.1.1

Sentence			% error in prediction of peak *P2*
(1)	b.	$\overline{E}$	3.94
		SD	4.21
(2)	b.	$\overline{E}$	5.87
		SD	7.80
(3)	b.	$\overline{E}$	−5.70
		SD	6.97
(4)	b.	$\overline{E}$	−2.33
		SD	2.48
Grand-mean	b.	$\overline{E}$	0.45
		SD	7.29

Table 2.10. Mean and grand-mean percent of error terms of the one predicted vs. obtained peak F_0 value in the group 2 (a) sentences of experiment 2.1.1

Sentence			% error in prediction of peak *P2*
(5)	a.	$\overline{E}$	7.21
		SD	3.88
(6)	a.	$\overline{E}$	−2.95
		SD	7.97
(7)	a.	$\overline{E}$	−2.71
		SD	8.54
(8)	a.	$\overline{E}$	5.68
		SD	7.78
Grand-mean	a.	$\overline{E}$	1.81
		SD	8.44

(b) sentences in Table 2.11. Grand-mean data and the associated Topline Rule for the respective groups above are found in Figures 2.13–2.16.

The fit between the predicted and obtained data for *P2* is very close in each case, with the obtained values in the short (a) sentences of both groups being slightly higher than the predicted values (Figures 2.13 and 2.15) and the obtained values in the long (b) versions of both groups being slightly lower than the predicted values (Figures 2.14 and 2.16).

The single-line version of theTopline Rule appears to model the average F_0 topline more than adequately for the data base considered thus far. Grand-

Table 2.11. Mean and grand-mean percent of error terms of the one predicted vs. obtained peak F_0 value in the group 2 (b) sentences of experiment 2.1.1

Sentence			% error in prediction of peak *P2*
(5)	b.	$\overline{E}$	4.67
		SD	8.67
(6)	b.	$\overline{E}$	−5.22
		SD	6.52
(7)	b.	$\overline{E}$	−10.49
		SD	4.98
(8)	b.	$\overline{E}$	4.37
		SD	7.42
Grand-mean	b.	$\overline{E}$	−1.67
		SD	9.40

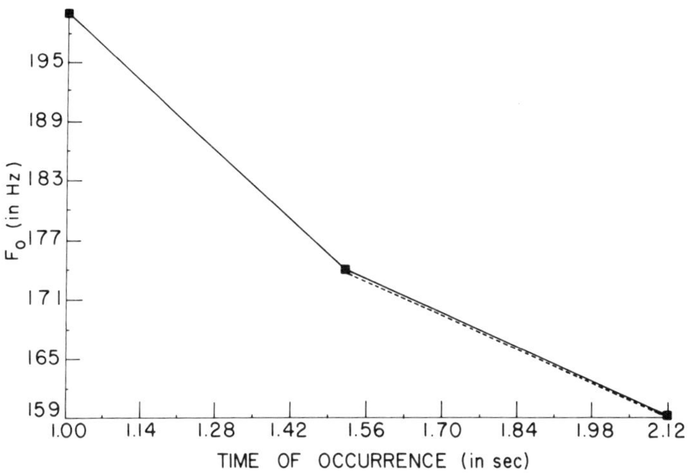

Figure 2.13. Predicted (dotted line) and obtained (solid line) values of mean F_0 vs. time for the grand mean data of Experiment 2.1.1, Group 1 (a) sentences, using Topline Model 3 to derive the predicted values.

mean error terms are typically near 1% or less for both male and female speakers. Examining the error terms by sex of speaker, we find that the grand-mean percent error for the twelve male speakers was 1.29% (SD 8.72) for *P2* and 0.97% (SD 5.23) for *P3*; for the eight females, the grand-mean percent error was −1.62% (SD 7.30) for *P2* and −1.06% (SD 4.98) for *P3*. Thus, the Topline Rule appears to capture the average form of peak declination in male and female speech with about the same high degree of accuracy.

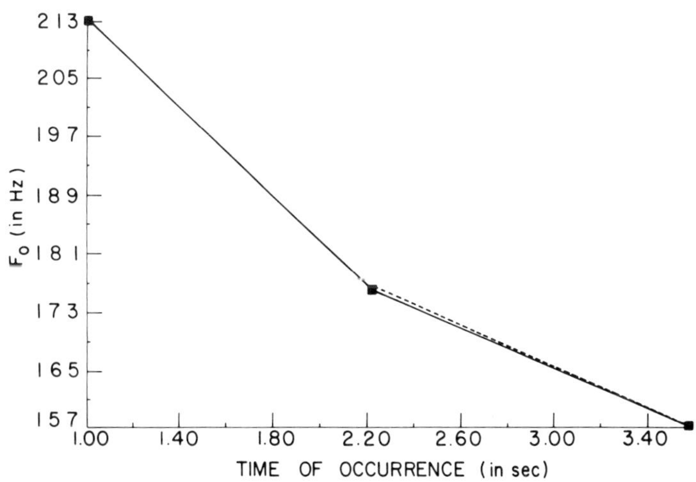

Figure 2.14. Predicted (dotted line) and obtained (solid line) values of mean F_0 vs. time for the grand mean data of Experiment 2.1.1, Group 1 (b) sentences, using Topline Model 3 to derive the predicted values.

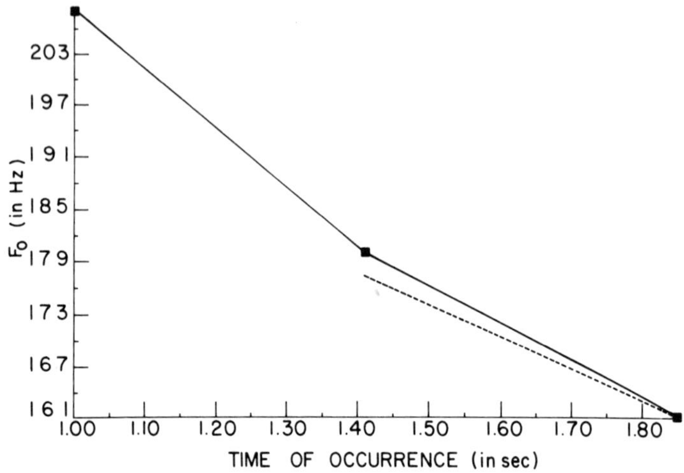

Figure 2.15. Predicted (dotted line) and obtained (solid line) values of mean F_0 vs. time for the grand mean data of Experiment 2.1.1, Group 2 (a) sentences, using Topline Model 3 to derive the predicted values.

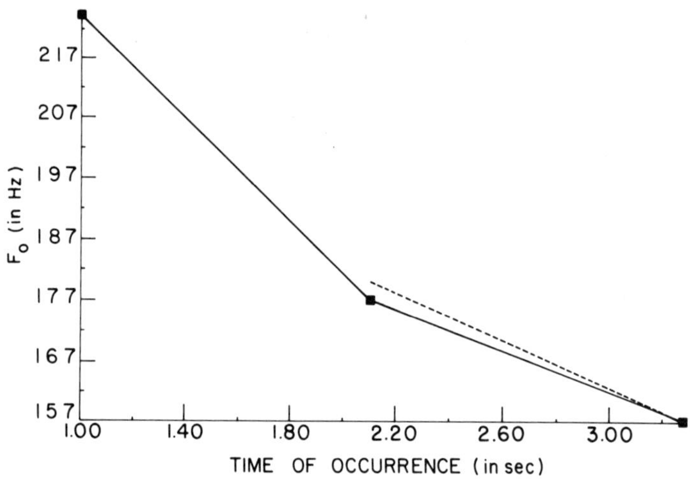

Figure 2.16. Predicted (dotted line) and obtained (solid line) values of mean F_0 vs. time for the grand mean data of Experiment 2.1.1, Group 2 (b) sentences, using Topline Model 3 to derive the predicted values.

Study 2.2. Topline Rule Invariance and Extensions

Having obtained a general formulation of the Topline Rule, it now seemed appropriate to rigorously test its predictive power in a variety of new circumstances that typify fluent speech, including variations in speaking rate and sentence structure. If the Topline Rule could adequately predict peak F_0 values despite wide variations in such factors, then this rule could be said to have indeed captured much of the essence of F_0 declination.

Experiment 2.2.1

In this experiment, we attempted to determine the effects of speaking rate on the F_0 topline. Nine speakers—seven females and two males—participated in this experiment. All nine served as subjects in Experiment 2.1.2. The sentence materials were identical to the set used in that experiment as well [Sentences (9)–(12)].

The subjects were not informed at the outset that they would be reading this list at different rates during the experimental session. The reading at a normal rate (Experiment 2.1.2) occurred first, usually as the second or third list in the session. After reading two intervening sentence lists on other issues, the subject was given the original list a second time and asked to read the sentences at a different rate, either slow or fast. After two or three more inter-

Table 2.12. Means and standard deviations of values of peak F_0 (Hz) and associated time of occurrence (sec) for each of the key words in the sentences spoken at normal rate in experiments 2.1.2 and 2.2.1

Sentence		Measurement			
		P1,T1	*P2,T2*	*P3,T3*	*P4,T4*
(9)	$\bar{F}_0$	197.	187.	175.	158.
	SD	40.5	43.6	44.3	43.4
	$\bar{t}$	1.00	1.47	2.06	2.68
	SD	0.00	0.05	0.15	0.18
(10)	$\bar{F}_0$	193.	178.	172.	166.
	SD	43.8	46.8	44.7	42.8
	$\bar{t}$	1.00	1.43	2.01	2.67
	SD	0.00	0.05	0.10	0.09
(11)	$\bar{F}_0$	219.	170.	170.	168.
	SD	54.1	43.1	43.6	46.1
	$\bar{t}$	1.00	1.47	2.04	2.49
	SD	0.00	0.08	0.10	0.12
(12)	$\bar{F}_0$	197.	190.	178.	168.
	SD	47.6	46.2	44.9	43.9
	$\bar{t}$	1.00	1.52	2.17	2.81
	SD	0.00	0.08	0.10	0.16
Grand-means and SDs	$\bar{F}_0$	201.	181.	174.	165.
	SD	45.9	43.7	42.6	42.3
	$\bar{t}$	1.00	1.47	2.07	2.66
	SD	0.00	0.07	0.12	0.18

Table 2.13. Means and standard deviations of values of peak F_0 (Hz) and associated time of occurrence (sec) for each of the key words in the sentences spoken at slow rate in experiment 2.2.1

Sentence		Measurement			
		P1,T1	*P2,T2*	*P3,T3*	*P4,T4*
(9)	$\bar{F}_0$	194.	178.	167.	156.
	SD	47.1	49.2	46.5	45.6
	$\bar{t}$	1.00	1.64	2.54	3.36
	SD	0.00	0.16	0.27	0.38
(10)	$\bar{F}_0$	181.	170.	166.	159.
	SD	42.4	45.1	43.9	46.1
	$\bar{t}$	1.00	1.63	2.53	3.36
	SD	0.00	0.28	0.45	0.47
(11)	$\bar{F}_0$	204.	163.	163.	163.
	SD	53.5	42.9	46.5	44.5
	$\bar{t}$	1.00	1.62	2.47	2.97
	SD	0.00	0.12	0.23	0.36
(12)	$\bar{F}_0$	191.	173.	170.	165.
	SD	52.4	52.6	50.8	45.9
	$\bar{t}$	1.00	1.71	2.59	3.39
	SD	0.00	0.16	0.32	0.43
Grand-means and SDs	$\bar{F}_0$	193.	171.	167.	161.
	SD	47.6	45.8	45.0	43.7
	$\bar{t}$	1.00	1.65	2.54	3.27
	SD	0.00	0.18	0.31	0.43

vening lists, the speaker was given the original list a third time and asked to read each sentence at the remaining rate. Five of the nine subjects read the list at a slow rate and later at a fast rate; the remaining four subjects performed these tasks in the opposite order. To alter the normal rate of speaking, the subject was instructed as follows: "Consider your normal rate of speaking, the rate which is normal for *you*, to have a magnitude of 100. I would like you to read the sentences here at a rate of 125 (fast rate) or 75 (slow rate). Try to keep your speech rate as consistent as possible from sentence to sentence." This procedure was adapted from Lane and Grosjean (1973) and has proved successful in this type of experiment. The utterances were then analyzed as previously described.

The data for the normal speaking rate were taken from Experiment 2.1.2, spoken by the same nine speakers who participated in the present experiment. The mean peak F_0 and time of occurrence for these data are found in Table 2.12. Table 2.13 contains the data spoken at the slow rate, and Table 2.14, at the fast rate. Figures 2.17–2.19 contain the mean peak F_0 data plotted against the mean time of occurrence for individual sentences at the normal, slow, and fast rates.

Table 2.14. Means and standard deviations of values of peak F_0 (Hz) and associated time of occurrence (sec) for each of the key words in the sentences spoken at fast rate in experiment 2.2.1

		Measurement			
Sentence		*P1,T1*	*P2,T2*	*P3,T3*	*P4,T4*
(9)	$\overline{F}_0$	201.	194.	178.	160.
	SD	52.2	48.8	44.2	41.0
	$\bar{t}$	1.00	1.41	1.91	2.41
	SD	0.00	0.04	0.10	0.19
(10)	$\overline{F}_0$	192.	189.	173.	165.
	SD	47.2	48.4	43.4	44.0
	$\bar{t}$	1.00	1.36	1.79	2.38
	SD	0.00	0.05	0.08	0.15
(11)	$\overline{F}_0$	214.	175.	173.	172.
	SD	57.6	43.2	44.0	45.6
	$\bar{t}$	1.00	1.44	1.85	2.25
	SD	0.00	0.06	0.09	0.12
(12)	$\overline{F}_0$	193.	189.	179.	171.
	SD	50.5	45.4	44.2	49.4
	$\bar{t}$	1.00	1.42	1.93	2.46
	SD	0.00	0.04	0.08	0.12
Grand-means and SDs	$\overline{F}_0$	200.	187.	176.	167.
	SD	50.5	45.0	42.1	43.4
	$\bar{t}$	1.00	1.41	1.87	2.38
	SD	0.00	0.06	0.10	0.16

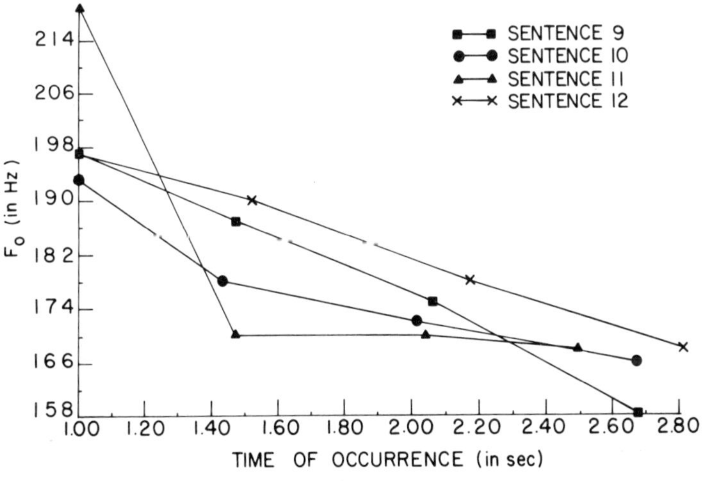

Figure 2.17. Plot of mean F_0 vs. time for each of the measured peak values in Sentences (9)–(12) spoken at normal rate in Experiment 2.2.1.

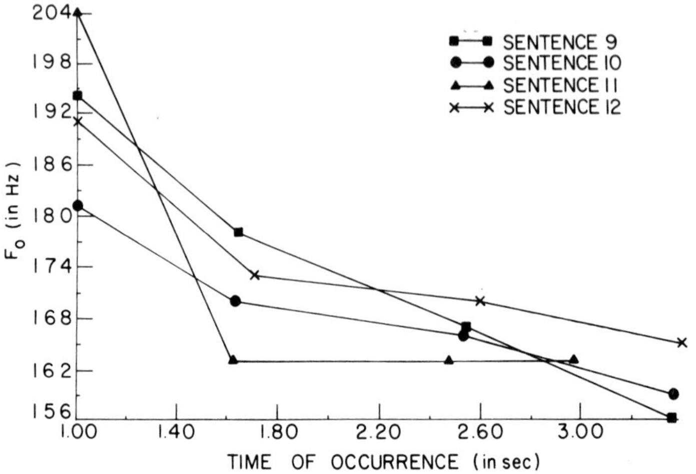

Figure 2.18. Plot of mean F_0 vs. time for each of the measured peak values in Sentences (9)–(12) spoken at slow rate in Experiment 2.2.1.

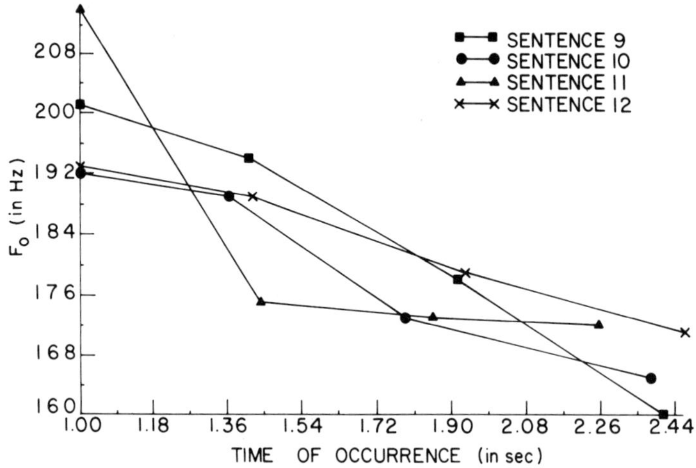

Figure 2.19. Plot of mean F_0 vs. time for each of the measured peak values in Sentences (9)–(12) spoken at fast rate in Experiment 2.2.1.

As in Experiment 2.1.2, the results for the individual sentences again show the influence of different inherent phonetic structure on F_0. In particular, the values of *P1* were systematically higher for Sentence (11) at each of the three speaking rates, while the values of *P2* for this same sentence were systematically lower than for the other sentences. As pointed out in Experiment 2.1.2, the high value of *P1* in (11) can be attributed to a combination of two facts: (a) the vowel /u/ in the key word *students* exhibits a higher inherent F_0 than any of the other vowels in *P1* words, and (b) the vowel /u/ is immediately preceded by a voiceless stop, also producing higher F_0 values during the vowel,

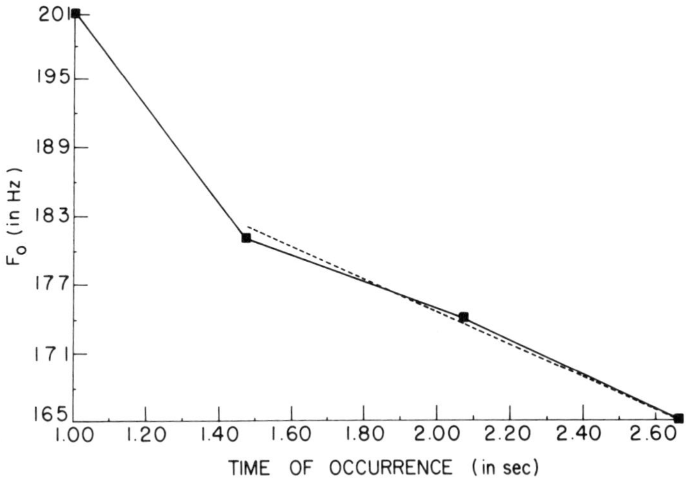

Figure 2.20. Predicted (dotted line) and obtained (solid line) values of mean F_0 vs. time for the grand mean data spoken at normal rate in Experiment 2.2.1.

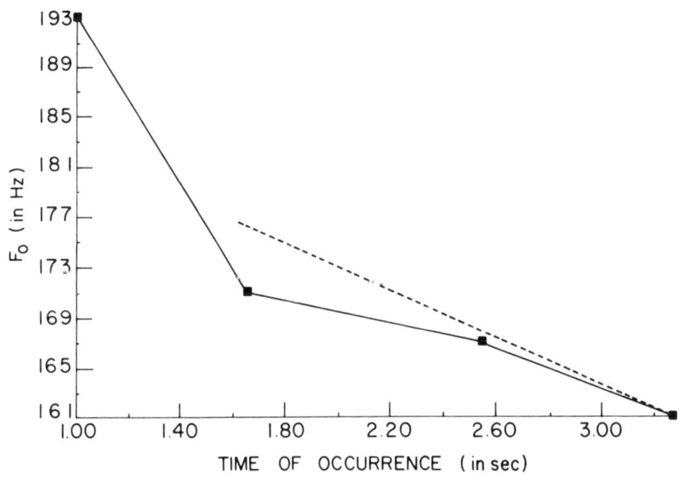

Figure 2.21. Predicted (dotted line) and obtained (solid line) values of mean F_0 vs. time for the grand mean data spoken at slow rate in Experiment 2.2.1.

while the low value of *P2* can be attributed, at least in part, to the fact that the key word *bus* begins with a voiced consonant, yielding relatively low F_0 values for the vowel. Figures 2.20–2.22 show the grand-mean peak F_0 and time of occurrence data at each rate along with the Topline Rule based on the grand-mean values of (*T1*,*P1*) and (*T4*,*P4*). Figure 2.23 shows the grand-mean data for all three rates together. Tables 2.15–2.17 contain the mean and grand-mean error terms based on applying the Topline Rule to individual utterances for the data of each of the three speaking rates.

The graph of the grand-mean data shown in Figure 2.23 reveals several

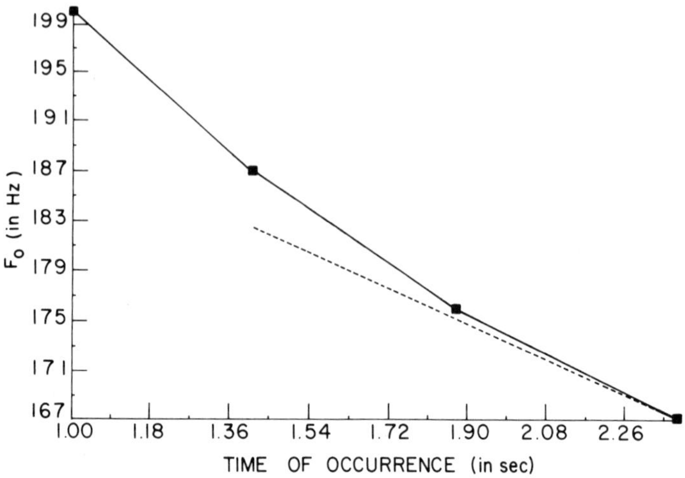

Figure 2.22. Predicted (dotted line) and obtained (solid line) values of mean F_0 vs. time for the grand mean data spoken at fast rate in Experiment 2.2.1.

interesting attributes of F_0 in speech spoken at different rates. To begin with, the mean peak F_0 given by *P2*, *P3*, and *P4* was lowest with the slow speech rate and highest for the fast rate. For *P1*, the slow rate exhibited the lowest mean F_0, but *P1* in the normal and fast rates were nearly identical, with the normal rate *P1* being 1 Hz higher than the fast rate *P1*. Table 2.18 contains the results of a series of t tests for matched pairs, comparing the peak F_0 value of each key word spoken at different rates. Such tests revealed that every mea-

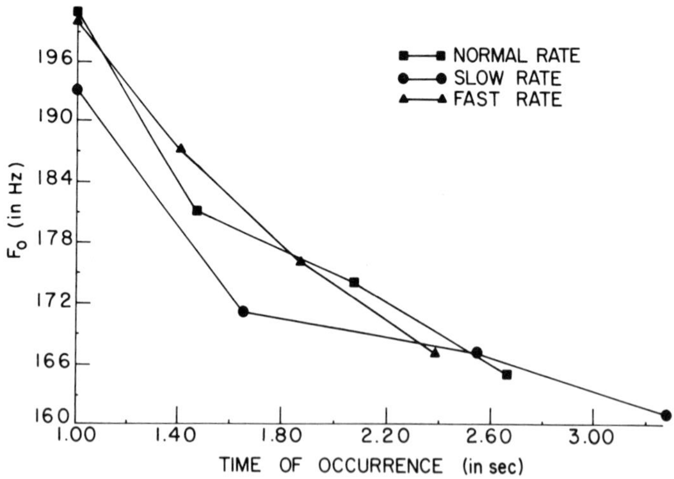

Figure. 2.23. Plot of mean F_0 vs. time for each of the measured peak values pooled across sentences at each rate in Experiment 2.2.1.

Table 2.15. Mean and grand-mean percent of error terms of the two predicted vs. obtained peak F_0 values in the sentences spoken at Normal rate in experiment 2.2.1

Sentence		% error in prediction of peak	
		P2	*P3*
(9)	$\overline{E}$	6.19	4.34
	SD	5.50	3.44
(10)	$\overline{E}$	−1.27	−0.41
	SD	6.23	4.13
(11)	$\overline{E}$	−10.94	−4.64
	SD	3.65	3.43
(12)	$\overline{E}$	4.73	2.27
	SD	6.88	3.96
Grand-means and SDs	$\overline{E}$	−0.32	0.39
	SD	8.74	4.95

Table 2.16. Mean and grand-mean percent of error terms of the two predicted vs. obtained peak F_0 values in the sentences spoken at slow rate in experiment 2.2.1

Sentence		% error in prediction of peak	
		P2	*P3*
(9)	$\overline{E}$	1.56	1.35
	SD	5.45	3.18
(10)	$\overline{E}$	−0.08	1.52
	SD	4.84	4.20
(11)	$\overline{E}$	−9.89	−4.21
	SD	3.79	6.31
(12)	$\overline{E}$	−3.22	−0.87
	SD	5.13	4.37
Grand-means and SDs	$\overline{E}$	−2.91	−0.55
	SD	6.42	5.03

Table 2.17. Mean and grand-mean percent of error terms of the two predicted vs. obtained peak F_0 values in the sentences spoken at fast rate in experiment 2.2.1

Sentence		% error in prediction of peak	
		P2	*P3*
(9)	$\overline{E}$	7.96	5.17
	SD	4.75	1.87
(10)	$\overline{E}$	5.74	0.37
	SD	3.98	3.13
(11)	$\overline{E}$	−7.56	−4.30
	SD	3.69	3.72
(12)	$\overline{E}$	5.00	2.23
	SD	6.08	5.22
Grand-means and SDs	$\overline{E}$	2.78	0.87
	SD	7.63	4.96

surement of peak F_0 at the slow rate was significantly lower than at the normal or fast rate. Between the normal and fast rates, there was also a significant difference in the values of *P2*. No other differences between the normal and fast rate were significant.

The finding of higher *P1* values at normal and fast rates of speech as opposed to slow rate can be compared with the findings for long vs. short utterances in Experiment 2.1.1. In that experiment, *P1* was significantly higher for longer utterances that contained more words and syllables than shorter utterances. This finding could in principle be attributed to the number of words or syllables in the utterance or to the anticipated time to speak the utterance. The results of this experiment indicate that the earlier effect is probably attributable to the number of words or syllables rather than to anticipated time. In this

Table 2.18. *t*-Tests of peak F_0 at one rate vs. another

	Measurement			
	P1	*P2*	*P3*	*P4*
Normal vs. slow	3.56[b]	4.77[c]	4.66[c]	2.24[a]
Fast vs. slow	2.16[a]	7.74[c]	5.45[c]	4.16[b]
Fast vs. normal	−.54	2.72[a]	1.52	1.81

[a] $p < .05$.
[b] $p < .002$.
[c] $p < .001$.

experiment, *P1* values for temporally long utterances at slow rate were actually lower than the *P1* values in temporally normal and short utterances (spoken at normal and fast rates, respectively).

In the present case, the consistently lower peak values obtained at slow rate may be attributed to a consequence of the speaker's lowered muscle tension. It is reasonable to assume that speakers exhibit less muscular tension in general when speaking slowly, including less tension of the vocal folds. As indicated in Chapter 1, slackened vocal folds contribute to a lowering of F_0. This slackening may be the explanation for the finding that each of the measured peaks contained a lower F_0 value at the slow rate of speaking.

The total sentence durations exhibited less difference between the fast and normal rate sentences than between the slow and normal rate sentences. The upper half of Table 2.19 shows the elapsed time from the mean of *T1* to the mean of the succeeding points in time for each of the three rates. The fast versions averaged 83% as long as the normal versions, while the slow versions averaged 137% as long as the normal versions. Thus, speakers did not speak as fast as the 75% instruction rate for fast speech but spoke slower than the 125% instruction rate for slow speech. The fact that the actual rate of the speakers departed from normal to a lesser extent at fast rates than at slow rates may be attributed to an articulatory constraint on the speaker's ability to compress the durations of segments at fast rates.

We turn now to a discussion of the accuracy of the Topline Rule at the different speaking rates. For the normal speaking rate, the remarkably close fit between the predicted and obtained values for both *P2* and *P3* in Figure 2.20 replicates the results shown earlier in Figure 2.12. Examination of Figures 2.21 and 2.22 shows that the average peak F_0 in the slow sentences appeared slightly below the Topline Rule, whereas the peaks in the fast sentences were displaced slightly above the Topline Rule, by about the same amount. These displacements are in agreement with the interpretation proposed above in terms of vocal fold tension. At fast rates, increased vocal fold tension gives rise to higher values of F_0, while at slow rates, decreased vocal fold tension yields lower values. The Topline Rule self-adjusts to these differences for the most part, as indicated by the relatively low error terms, yet the differences between the predicted and obtained peak values indicate that these adjustments are slightly less than adequate.

Table 2.19. Elapsed time (sec) for grand-mean data of experiment 2.2.1

	Time interval			
Rate	*T2–T1*	*T3–T2*	*T4–T3*	Mean
Fast	.41	.46	.51	.46
Normal	.47	.60	.59	.55
Slow	.65	.89	.73	.76

Experiment 2.2.2

This experiment was designed to test whether the presence of a subordinate restrictive relative clause could perturb the F_0 topline. Three test sentences were constructed for the experiment, each containing a subordinate relative clause beginning with the relative pronoun *who.*

The sentence materials appear below, with the key words underlined:

(13) *My cousin met the man who was buying milk and bread.*
(14) *The man who your cousin met was buying milk and bread.*
(15) *The man who met your cousin was buying milk and bread.*

In order to examine the generality of the Topline Rule for different types of restrictive relative clauses, these sentences differed from one another in the grammatical relations of the relative and main clauses. In (13), the relative clause modifies the direct object of the main clause, while in (14) and (15), the relative clause modifies the main clause subject. The latter two sentences differ from one another in terms of the underlying grammatical relation of the *wh* word within the relative clause itself. In (14), this word appears as an underlying direct object, whereas in (15), it appears as an underlying subject. In each sentence, key words were measured in both the main and relative clauses in order to assess the generality of the Topline Rule in such complex structures.

Ten subjects, four males and six females, all of whom had participated in Experiment 2.1.2, read these experimental sentences. In each sentence, the peak F_0 and time of occurrence were obtained for each of the six underlined key words. In addition to their structural similarity, all three sentences contained the same key words. The last three key words appeared in corresponding positions among the three sentences, and the first key word of the latter two sentences was also identical. Thus, we were able to compare particular peak F_0 values of each sentence against those in other sentences in a controlled fashion.

The mean peak F_0 values and times of occurrence are given in Table 2.20. The data for the individual sentences are plotted in Figure 2.24; the grand-mean data and the associated Topline Rule are shown in Figure 2.25. Finally, the mean percent error terms are given in Table 2.21. Within the final phrase, . . . *was buying milk and bread,* the mean values for the three F_0 peaks were very similar across the three test sentences; the largest range was 2 Hz for *P4*. These results suggest that the F_0 peaks in restrictive relative clauses do not differ substantially from those in comparable positions within main clauses.

The peak data for the first three peaks, however, revealed apparent new topline anomalies. These data showed a strong similarity between the mean values of *P1* and *P2* among the three sentences. Previous experiments yielded *P2* values that were considerably lower than *P1*, but such was not the case here. In addition, the mean *P3* values in Sentences (13) and (14) were considerably lower than their corresponding *P2* values. However, in Sentence (15), the mean *P3* value was only slightly lower than the previous *P2* value.

It seems that each of these anomalies may be due to independent local influ-

Table 2.20. Means and standard deviations of values of peak F_0 (Hz) and associated time of occurrence (sec) for each of the key words in the sentences of experiment 2.2.2

Sentence		Measurement					
		P1,T1	*P2,T2*	*P3,T3*	*P4,T4*	*P5,T5*	*P6,T6*
(13)	$\bar{F}_0$	186.	179.	163.	158.	157.	153.
	SD	51.4	53.7	45.8	47.0	44.5	43.2
	$\bar{t}$	1.00	1.31	1.64	2.24	2.57	2.94
	SD	0.00	0.04	0.08	0.12	0.15	0.19
(14)	$\bar{F}_0$	183.	182.	166.	158.	158.	154.
	SD	53.0	53.6	50.7	46.3	45.4	43.4
	$\bar{t}$	1.00	1.51	1.82	2.37	2.69	3.05
	SD	0.00	0.11	0.12	0.22	0.21	0.26
(15)	$\bar{F}_0$	186.	180.	178.	156.	157.	153.
	SD	50.3	51.6	49.1	46.4	45.3	44.4
	$\bar{t}$	1.00	1.26	1.62	2.20	2.49	2.86
	SD	0.00	0.04	0.05	0.20	0.14	0.17
Grand-means and SDs	$\bar{F}_0$	185.	180.	169.	157.	157.	153.
	SD	49.8	51.2	47.3	45.0	43.5	42.1
	$\bar{t}$	1.00	1.36	1.69	2.27	2.58	2.95
	SD	0.00	0.13	0.12	0.19	0.19	0.22

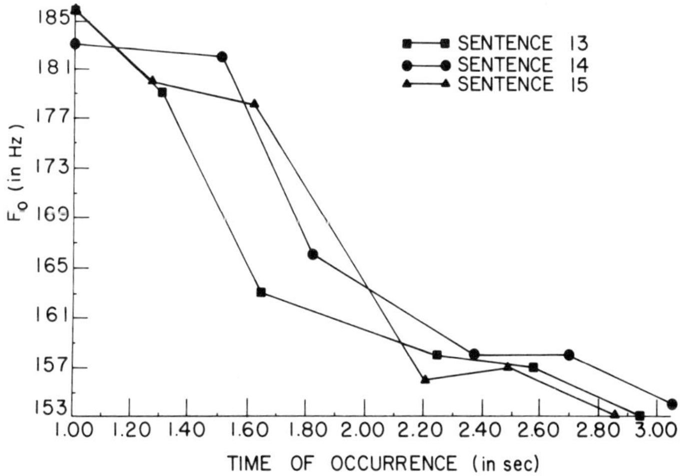

Figure 2.24. Plot of mean F_0 vs. time for each of the measured peak values in Sentences (13)–(15).

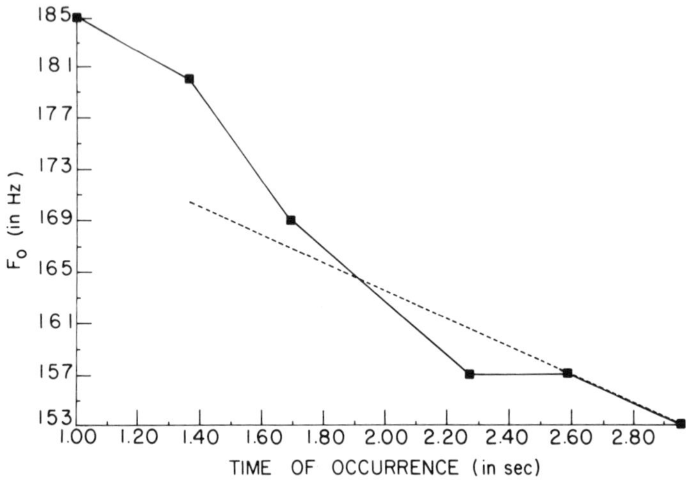

Figure 2.25. Predicted (dotted line) and obtained (solid line) values of mean F_0 vs. time for the pooled data of Experiment 2.2.2.

ences. Note that the key word *cousin,* no matter where it appears, shows a high mean F_0. In these particular test sentences, *cousin* may receive extra focus, carrying more semantic load than *man* or *met.* Also, *cousin* begins with the voiceless stop /k/, which tends to raise F_0 for the first 100 msec or so of the succeeding vowel. These influences may account for the high mean value of *P2* in Sentence (14) and *P3* in Sentence (15). We must now account for the high values of *P2* in Sentences (13) and (15). In Sentence (15), the *P2* word *(met)* is surrounded by a high *P1* word *(man)* and a high *P3* word *(cousin).* We suggest that the speaker finds it easier to raise F_0 on the *P2* word to the approximate range of the *P1* and *P3* words than to execute a sharp fall and rise in F_0 from *P1* to *P2* to *P3.* This suggestion of forwards coarticulation implies that

Table 2.21. Mean and grand-mean percent of error terms of the four predicted peak F_0 values in the sentences of experiment 2.2.2

		% error in prediction of peak			
Sentence		*P2*	*P3*	*P4*	*P5*
(13)	$\overline{E}$	3.62	−2.95	−2.20	−0.12
	SD	4.74	4.82	2.82	2.39
(14)	$\overline{E}$	7.98	0.10	−1.13	0.43
	SD	5.51	7.44	4.03	2.94
(15)	$\overline{E}$	4.49	6.49	−3.11	0.14
	SD	4.90	6.91	4.88	3.23
Grand-means and SDs	$\overline{E}$	5.36	1.21	−2.15	0.15
	SD	5.25	7.43	3.95	2.78

the speaker has looked ahead and already formulated some kind of idea about the F_0 contour; that he "knows" the *P3* word will have a high F_0, and adjusts the *P2* value accordingly.

Finally, what about the high value of *P2* in Sentence (13)? Here we postulate that the extremely short temporal interval (260 msec on the average) between the location of *P1* and *P2* seems an insufficient amount of time for F_0 to fall from *P1* to the much lower values for *P2* that we have observed in previous experiments. This short interval may also play a part in keeping the *P2* value so high in Sentence (15).

As one might expect, the high *P2* values are not predicted particularly well by the Topline Rule. Over all utterances, the mean error for the *P2* prediction, referring to Table 2.21, was 5.36%, while the terms for the remaining peaks were much smaller in magnitude. In particular, the standard deviations of the error terms for *P4* and *P5* were quite small, confirming other data which indicate only small perturbations in these peak values across sentences.

In summary, the results of this experiment indicate that the Topline Rule can adequately predict the F_0 peaks in sentences containing a restrictive relative clause, including both peaks within the main and relative clauses. While minor deviations from this rule were observed, these perturbations seemed attributable to independent, local factors. Further work must be conducted, however, before we can ascertain whether this account is correct or whether we can have simply been guilty of attempting to "explain away" the perturbations in an unprincipled fashion, weakening the falsifiability of the Topline Rule. For the remaining peaks, the fit between the predicted and obtained values was very close, providing further support for the predictive power of the Topline Rule across a variety of different sentence types.

Experiment 2.2.3

As in the previous experiment, this test was designed to assess whether the inclusion of a subordinate clause would influence the form of the F_0 topline. In this case, the test sentences contained complement clauses, beginning with the complementizer *that.* Complement clauses differ from restrictive relatives in that complements appear within the scope of a superordinate Verb. While relative clauses strictly modify a superordinate Noun Phrase that appears in surface structure, complements may serve as substitutes for superordinate Noun Phrases or Verb Phrases.

The sentences for this experiment are shown below, with the key words underlined:

(16) *The <u>coach</u> <u>says</u> that <u>eating</u> <u>carrots</u> <u>improves</u> your <u>game</u>.*

(17) *The <u>doctor</u> <u>suspects</u> that the <u>man</u> next <u>door</u> is <u>very</u> <u>sick</u>.*

The subject pool for this experiment was the same as that of Experiment 2.2.2. The peak F_0 and time of occurrence was measured for each key word in each

Table 2.22. Means and standard deviations of values of peak F_0 (Hz) and associated time of occurrence (sec) for each of the key words in the sentences of experiment 2.2.3

Sentence		Measurement					
		P1,T1	*P2,T2*	*P3,T3*	*P4,T4*	*P5,T5*	*P6,T6*
(16)	$\bar{F}_0$	196.	177.	175.	172.	166.	157.
	SD	59.5	53.8	53.7	48.6	45.9	43.7
	$\bar{t}$	1.00	1.29	1.73	2.05	2.61	3.04
	SD	0.00	0.05	0.12	0.13	0.17	0.20
(17)	$\bar{F}_0$	183.	169.	163.	161.	154.	149.
	SD	49.2	44.6	47.0	47.4	43.3	43.9
	$\bar{t}$	1.00	1.59	2.25	2.69	3.10	3.44
	SD	0.00	0.08	0.17	0.19	0.31	0.30
Grand-means and SDs	$\bar{F}_0$	190.	173.	169.	166.	160.	153.
	SD	53.6	48.3	49.4	47.1	43.8	42.9
	$\bar{t}$	1.00	1.44	1.99	2.37	2.86	3.24
	SD	0.00	0.17	0.30	0.36	0.35	0.32

utterance. In Table 2.22, we present the mean peak F_0 and time of occurrence for each measurement, and in Figure 2.26, a graph of the data for individual sentences. Figure 2.27 shows a plot of the grand-mean data and the corresponding Topline Rule. Table 2.23 contains the mean error terms based on Topline Rule predictions of *P2* through *P5*.

One difference between the two sentences was exhibited in the amount of time taken to speak them. Note that the amount of time between *P1* and *P2*

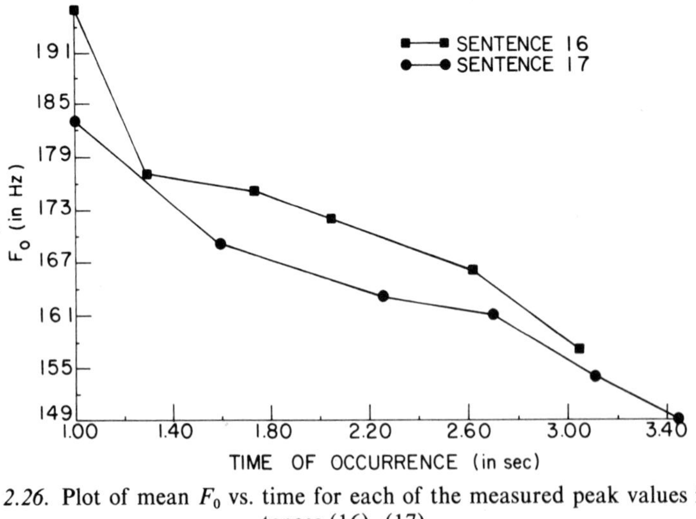

Figure 2.26. Plot of mean F_0 vs. time for each of the measured peak values in Sentences (16)–(17).

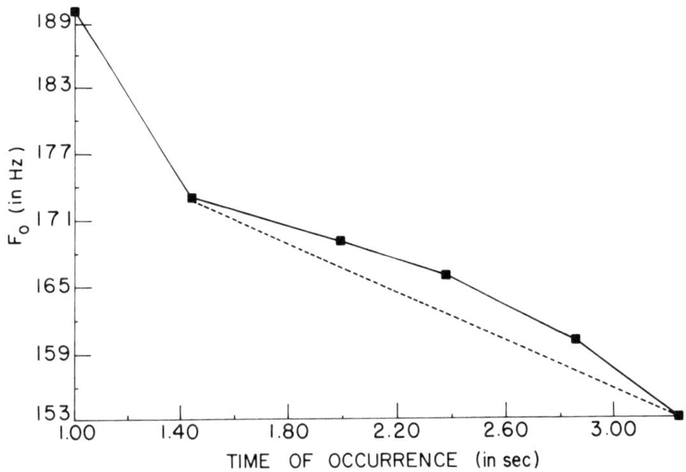

Figure 2.27. Predicted (dotted line) and obtained (solid line) values of mean F_0 vs. time for the pooled data of Experiment 2.2.3.

in Sentence (16) averaged 290 msec, while in Sentence (17), this difference averaged 590 msec. The graph of the individual sentence data in Figure 2.26 highlights these differences. Overall, the mean duration between *T1* and *T6* was 2.04 sec in Sentence (16) and 2.44 sec in Sentence (17). Despite differences in the stress pattern of words and number of syllables between measurements in these two sentences, the Topline Rule models the peak F_0 values in both sentences fairly accurately, as seen in Table 2.23 and Figure 2.27, with the obtained peak values lying very slightly above the predicted values. As indicated in the table, the largest grand-mean error occurs in the prediction of *P4*, 2.36%. The remaining grand-mean percent error terms range from about .5% to about 2%, and all exhibit standard deviations of about 5%. These low error terms suggest that the inclusion of a complement clause does not significantly alter the topline.

Table 2.23. Mean and grand-mean percent of error terms of the four predicted vs. obtained peak F_0 values in the sentences of experiment 2.2.3

Sentence		% error in prediction of peak			
		P2	*P3*	*P4*	*P5*
(16)	$\overline{E}$	−1.42	−0.04	1.50	2.36
	SD	5.76	4.42	3.54	3.00
(17)	$\overline{E}$	2.37	2.13	3.21	1.95
	SD	3.75	5.49	6.42	6.37
Grand-means and SDs	$\overline{E}$	0.48	1.04	2.36	2.15
	SD	5.11	4.98	5.12	4.85

Experiment 2.2.4

Until now, the key words measured in each of the experiments have consisted of Nouns and Verbs. In this experiment, the measurements of peak F_0 were applied to two other major grammatical categories, Adjectives and Adverbs, as a means of further testing the predictive power of the Topline Rule.

Two groups of test sentences were prepared for this experiment. In Group 1, each sentence contained six key words; in Group 2, five key words. The data for each group will be presented separately. The same ten subjects of Experiment 2.2.2 participated here. The test sentences of Group 1 are shown below, with the key words underlined:

(18) *Six artists carefully sketched the forest scene.*
(19) *Brick houses are often found in classy suburbs.*

Table 2.24 shows the mean peak F_0 and time of occurrence for each of the six key words in Group 1. These data are presented graphically in Figure 2.28.

The major difference in the F_0 values of the two sentences involved *P3*, in which a substantially higher value was obtained in Sentence (18) than in (19). The higher *P3* value in (18) may be attributed to the possibility that the Adverb *carefully* carries greater information load (and hence more inherent stress; Lightfoot, 1970; Coker, Umeda, and Browman, 1973) than the adverb *often* in (19). The grand-mean data and Topline Rule shown in Figure 2.29 and the error terms of Table 2.25 reveal a fairly good fit with the Topline Rule; the largest grand-mean error term was −3.73% for *P4*. Three of the four

Table 2.24. Means and standard deviations of values of peak F_0 (Hz) and associated time of occurrence (sec) for each of the key words in the group 1 sentences of experiment 2.2.4

Sentence		Measurement					
		P1,T1	*P2,T2*	*P3,T3*	*P4,T4*	*P5,T5*	*P6,T6*
(18)	$\bar{F}_0$	207.	180.	181.	164.	160.	155.
	SD	53.6	47.8	44.3	45.4	41.3	46.9
	$\bar{t}$	1.00	1.31	1.84	2.38	2.89	3.29
	SD	0.00	0.06	0.18	0.20	0.27	0.29
(19)	$\bar{F}_0$	200.	188.	165.	163.	159.	159.
	SD	55.4	48.4	43.8	46.7	44.7	43.0
	$\bar{t}$	1.00	1.20	1.79	2.17	2.68	3.09
	SD	0.00	0.04	0.13	0.13	0.21	0.23
Grand-means and SDs	$\bar{F}_0$	204.	184.	173.	164.	160.	157.
	SD	53.2	47.0	43.6	44.8	41.9	43.8
	$\bar{t}$	1.00	1.26	1.81	2.28	2.78	3.19
	SD	0.00	0.08	0.15	0.20	0.26	0.28

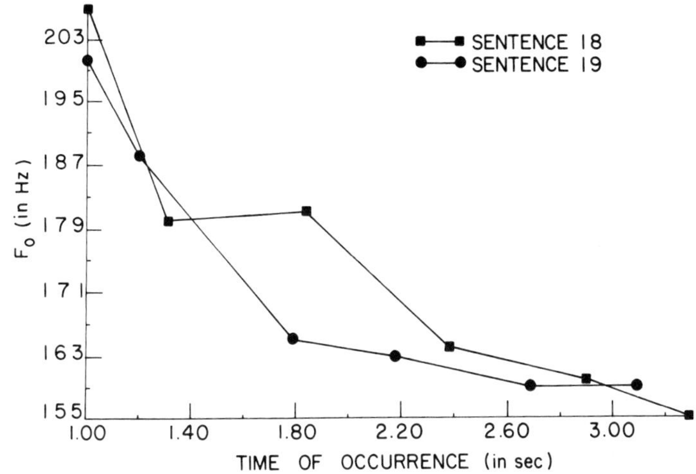

Figure 2.28. Plot of mean F_0 vs. time for each of the measured peak values in Sentences (18)–(19).

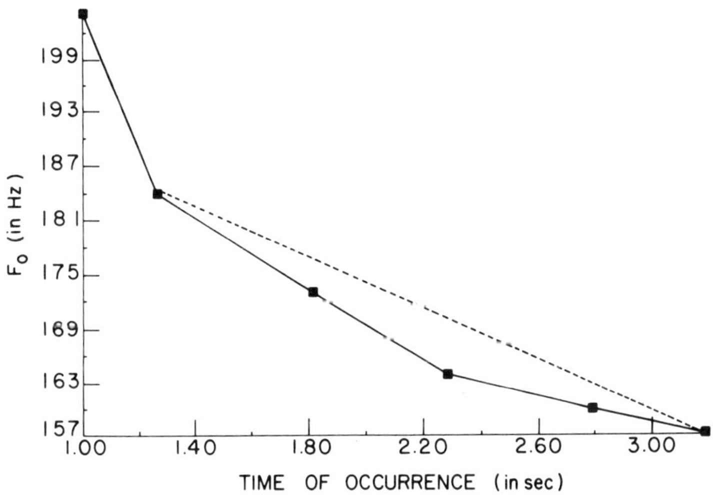

Figure 2.29. Predicted (dotted line) and obtained (solid line) values of mean F_0 vs. time for the pooled data of Experiment 2.2.4.

Table 2.25. Mean and grand-mean percent of error terms of the four predicted vs. obtained peak F_0 values in the group 1 sentences of experiment 2.2.4

		% error in prediction of peak			
Sentence		*P2*	*P3*	*P4*	*P5*
(18)	$\overline{E}$	−2.69	2.96	−2.28	0.56
	SD	6.08	7.39	4.89	7.37
(19)	$\overline{E}$	2.80	−6.42	−5.18	−3.42
	SD	6.34	4.36	5.92	2.83
Grand-means and SDs	$\overline{E}$	0.05	−1.73	−3.73	−1.43
	SD	6.67	7.62	5.49	5.80

grand-mean error terms were negative, and the one positive term was just barely so (.05% for *P2*). The overall negative error terms may be attributed to some amount of extra stress placed on the first key word in each sentence. Both initial key words were utterance-initial, stressed, monosyllabic Adjectives. A higher than normal value of *P1* would shift the Topline Rule upward, thus causing the predicted values of F_0 to lie above the actual values, and producing error terms with negative sign, as obtained here.

The two sentences in Group 2 also contained a variety of grammatical categories among the key words:

(20) *The cat in the garage ran swiftly underneath the car.*
(21) *Days pass quickly in the middle of winter.*

Unlike Group 1, these sentences contained a total of five key words, yielding three utterance-internal peaks in each sentence to be predicted using the Topline Rule. The mean peak F_0 and times of occurrence are given in Table 2.26; the data are graphed in Figure 2.30.

Considering first the plots of the individual sentence data in Figure 2.30, one notes a significant departure from the "normal" F_0 topline pattern in Sentence (20). The successive error terms for this one sentence, found in Table 2.27, were approximately −1% for *P2*, +8% for *P3*, and − 3.5% for *P4*. The general pattern of F_0 declination was evident for this sentence, but why was the F_0 topline such a zigzag? It seems likely that the *P3* key word, *swiftly,* was responsible for the perturbation. The error term for *P3* in Sentence (20) was

Table 2.26. Means and standard deviations of values of peak F_0 (Hz) and associated time of occurrence (sec) for each of the key words in the group 2 sentences of experiment 2.2.4

Sentence		Measurement				
		P1,T1	*P2,T2*	*P3,T3*	*P4,T4*	*P5,T5*
(20)	$\bar{F}_0$	197.	178.	183.	161.	159.
	SD	49.9	50.7	42.9	43.9	40.2
	$\bar{t}$	1.00	1.46	2.17	2.50	3.14
	SD	0.00	0.11	0.10	0.10	0.19
(21)	$\bar{F}_0$	194.	194.	185.	160.	159.
	SD	49.8	52.1	42.1	44.5	43.8
	$\bar{t}$	1.00	1.22	1.58	2.12	2.49
	SD	0.00	0.09	0.08	0.17	0.17
Grand-means and SDs	$\bar{F}_0$	195.	186.	184.	160.	159.
	SD	48.6	50.7	41.4	43.1	40.9
	$\bar{t}$	1.00	1.34	1.87	2.31	2.81
	SD	0.00	0.16	0.32	0.24	0.38

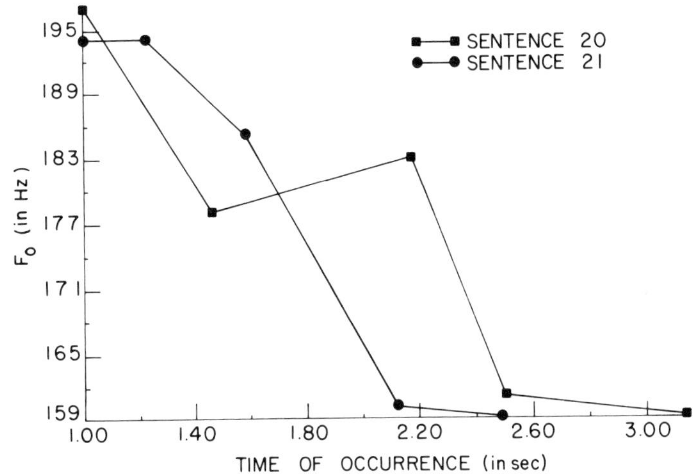

Figure 2.30. Plot of mean F_0 vs. time for each of the measured peak values in Sentences (20)–(21).

by far the largest, nearly 8%. Since this word is an Adverb, it probably received more stress than a non-Adverb at the same sentence location (Lightfoot, 1970; O'Shaughnessy, 1976). In addition, it seems reasonable that the extra focus given to the Adverb might cause a defocusing of the neighboring key words, particularly the following word, which could explain why the mean error term for *P4* was negative (see also Chapter 4). In short, the focused Adverb pulls up on the topline; to compensate, a lowering of the topline occurs just after the focus, creating the observed zigzag pattern. The present rationale is admittedly ad hoc, but such proposals seem useful at this rudimentary stage of F_0 research in sentence contexts, in order to suggest directions for independent further testing.

Table 2.27. Mean and grand-mean percent of error terms of the three predicted vs. obtained F_0 values in the group 2 sentences of experiment 2.2.4

		% error in prediction of peak		
Sentence		*P2*	*P3*	*P4*
(20)	$\bar{E}$	−0.92	7.89	−3.60
	SD	5.48	7.12	5.20
(21)	$\bar{E}$	8.86	8.17	−3.40
	SD	5.90	9.57	3.60
Grand-means and SDs	$\bar{E}$	3.97	8.03	−3.50
	SD	7.47	8.21	4.35

The mean error terms of Sentence (21) show essentially the same features as Sentence (20), with the one exception that the mean percent error for *P2* was much higher. This exception may be attributed to the short mean duration of 220 msec between *T1* and *T2* in Sentence (21), similar to the situation observed in Experiment 2.3.2. In Sentence (20), the mean interval between *T1* and *T2* was over twice as long, 460 msec (reading directly from Table 2.26). In a sentence such as *Days pass quickly* . . . very little time was available for executing large F_0 transitions from the first to second syllable, and consequently, the mean peak F_0 from *P1* to *P2* was about the same.

In summary, the data for both sentences indicate that Adverbs typically receive more inherent stress than words belonging to other grammatical categories. This result is explicable in terms of the fact that Adverbs often appear as focused elements in the sentence.

Experiment 2.2.5

In a further attempt to test perturbations of the F_0 topline, sentences were constructed that contained parenthetical clauses. Parentheticals are typically demarcated from the main clause by pauses (e.g., Downing, 1973; Emonds, 1973, 1976). In addition, such expressions seem to be produced with relatively low stress, accompanied by low values of both F_0 and amplitude in comparison to the prosodic attributes of the main clause. Given these characteristics, there is little doubt that a single topline function cannot adequately capture the F_0 peaks contained in both main and parenthetical clauses within the same utterance. Since the Topline Rule was originally designed to account for F_0 peaks in main clauses, we tested its generality by asking whether the presence of parenthetical clauses would disrupt the topline within the main clause.

This experiment contained two different groups of experimental sentences. During the original experimental sessions, Group 1 was included in a single list and Group 2 was contained in three separate lists.

Group 1

The ten subjects who took part in this section of the experiment also participated in Experiment 2.1.2. The present group of sentences were, in fact, similar to the sentences of that experiment. Our technique involved inserting a parenthetical clause *(it seems to me)* into these sentences, whose F_0 topline had already been accurately modeled by the Topline Rule. The sentences, with underlined key words, were as follows:

(22) *The book on the table, it seems to me, was a gift from my mother.*
(23) *The man in the truck, it seems to me, sent roses to his niece.*
(24) *The students on the bus, it seems to me, threw gum at the teacher.*
(25) *The ducks in the park, it seems to me, will eat popcorn and peanuts.*

Eight matched filler sentences accompanied the four test sentences. No commas appeared on the typed sentence lists to set off the parenthetical clauses. The subjects were instructed that the clause *it seems to me* was to be considered a parenthetical, and the experimenter spoke several examples of other types of sentences with parenthetical clauses, in order to familiarize the subject with this structure.

The peak F_0 and time of occurrence was measured for each of the six key words in each test sentence. The mean and standard deviations for these data are shown in Table 2.28. The mean data for individual sentences are shown graphically in Figure 2.31 and the mean over the entire group of four sentences with the Topline Rule in Figure 2.32. Mean error terms for individual sentences and across all sentences are shown in Table 2.29.

Let us first consider the data and plots of the four individual sentences. Figure 2.31 indicates that the peak F_0 dropped off sharply from *P2* to *P3* and rose again from *P4* to *P5*. The mean values of *P3* and *P4*, measured inside the parenthetical, were markedly lower than the mean peak F_0 of the words on

Table 2.28. Means and standard deviations of values of peak F_0 (Hz) and associated time of occurrence (sec) for each of the key words in the sentences of experiment 2.2.5

Sentence		Measurement					
		P1,T1	*P2,T2*	*P3,T3*	*P4,T4*	*P5,T5*	*P6,T6*
(22)	$\bar{F}_0$	202.	194.	161.	157.	172.	156.
	SD	50.7	49.0	45.2	47.4	46.1	40.6
	$\bar{t}$	1.00	1.48	2.06	2.42	3.10	3.70
	SD	0.00	0.05	0.11	0.13	0.21	0.27
(23)	$\bar{F}_0$	189.	188.	163.	156.	169.	161.
	SD	45.8	45.4	43.7	45.3	43.6	45.0
	$\bar{t}$	1.00	1.39	1.93	2.31	2.96	3.67
	SD	0.00	0.07	0.16	0.17	0.33	0.34
(24)	$\bar{F}_0$	211.	179.	158.	151.	167.	169.
	SD	54.0	46.3	43.0	44.4	45.5	47.8
	$\bar{t}$	1.00	1.50	2.13	2.48	3.17	3.62
	SD	0.00	0.05	0.17	0.17	0.31	0.33
(25)	$\bar{F}_0$	191.	190.	163.	159.	172.	168.
	SD	47.1	48.8	48.2	48.3	45.6	43.6
	$\bar{t}$	1.00	1.51	2.05	2.43	3.11	3.73
	SD	0.00	0.07	0.12	0.12	0.24	0.28
Grand-means and SDs	$\bar{F}_0$	199.	188.	161.	156.	170.	163.
	SD	48.4	45.9	43.4	44.6	43.5	42.9
	$\bar{t}$	1.00	1.47	2.04	2.41	3.08	3.68
	SD	0.00	0.08	0.15	0.16	0.28	0.30

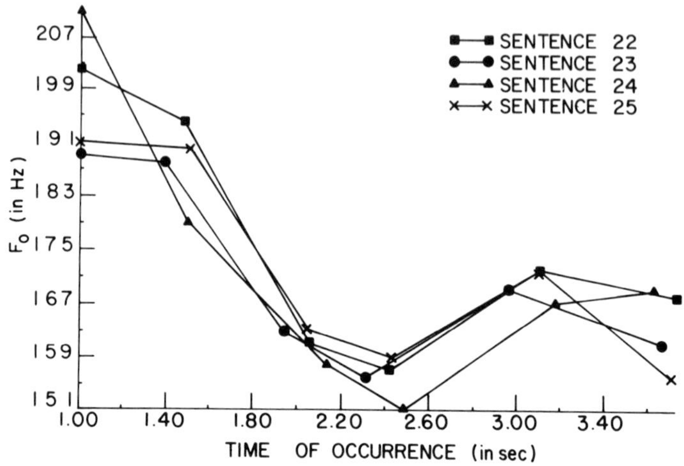

Figure 2.31. Plot of mean F_0 vs. time for each of the measured peak values in Sentences (22)–(25).

either side of the parenthetical. The plot of the grand mean data and associated Topline Rule in Figure 2.32 also indicates that the peak F_0 values inside the parenthetical were much lower than in the surrounding clause. These results are in full accord with the intuition stated at the outset of this study regarding the lowered stress generally accompanying parenthetical material. The Topline Rule based on the main clause, of course, lies well above the actual values of the peaks *within* the parenthetical. Since only two peaks were measured within the parenthetical, it was not possible to generate a separate Topline Rule based on this material alone, but in future work it would be interesting to determine

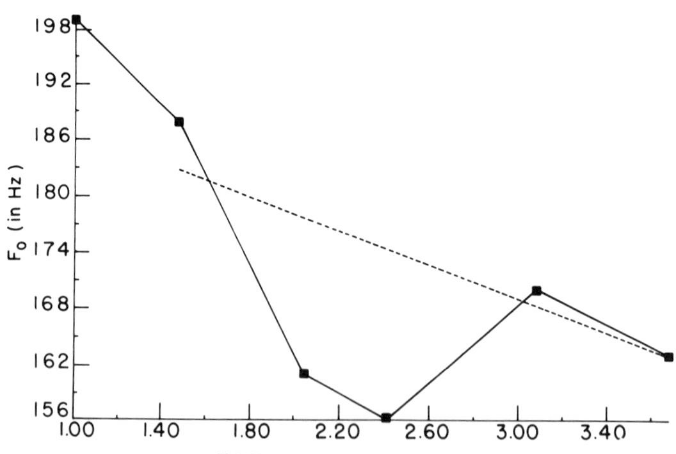

Figure 2.32. Predicted (dotted line) and obtained (solid line) values of mean F_0 vs. time for the pooled data of Sentences (22)–(25).

Table 2.29. Mean and grand-mean percent of error terms of the four predicted vs. obtained peak F_0 values in the sentences of experiment 2.2.5

Sentence		% error in prediction of peak			
		P2	*P3*	*P4*	*P5*
(22)	$\overline{E}$	7.25	−8.37	−9.19	5.41
	SD	6.89	6.22	7.40	4.00
(23)	$\overline{E}$	6.64	−6.25	−9.22	2.00
	SD	5.52	5.68	5.93	3.38
(24)	$\overline{E}$	−6.17	−14.17	−16.68	−3.75
	SD	7.23	6.09	5.66	4.84
(25)	$\overline{E}$	5.17	−9.15	−10.31	0.08
	SD	4.28	7.29	7.78	6.78
Grand-means and SDs	$\overline{E}$	3.22	−9.48	−11.35	0.93
	SD	8.06	6.77	7.21	5.79

whether this material adheres to its own version of a Topline Rule, in which the domain of the rule includes only the parenthetical material itself. Figure 2.32 also shows that the Topline Rule provides a fairly good description of the peaks contained within the MAIN clause (*P2* and *P5*) despite the presence of the parenthetical, although the values for *P2* were somewhat higher than predicted (see below).

At this point, it is useful to provide a comparison of the present data with that from Experiment 2.1.2 for the same ten subjects who participated in this experiment. In Table 2.30, we present a measurement-by-measurement comparison of the data for matched key words from Experiment 2.1.2 vs. the data of the present experiment for these ten speakers. The mean data are graphed in Figure 2.33. *t* tests for matched pairs were performed on these data in order to assess any significant differences. The only consistent F_0 difference involved significantly higher values of *P2* in three of the four sentences, (22)–(24) ($p < .05$, t values ranging from 2.55 to 3.60, $df = 9$).

The consistently higher values of *P2* in the sentence with parentheticals seem to reflect a heightening of stress on the word just prior to the beginning of the parenthetical expression. Although the valley of the *P2* word was not measured here, it seems that a somewhat exaggerated fall in F_0 also accompanies the word immediately preceding a parenthetical, as a means of demarcating the beginning of the parenthetical from the main clause.

It is noteworthy that no consistent differences were obtained for the values of *P1* in these comparisons, even though the sentences with parentheticals were somewhat longer than those without parentheticals. This finding contrasts with an earlier finding in Experiment 2.1.1, showing *P1* higher in longer utterances (also obtained by McAllister, 1971, and O'Shaughnessy, 1976). The absence of a consistent effect here may be attributed to one of two factors: (a) the

Table 2.30. Comparison of mean peak F_0 (Hz) and time of occurrence (sec) of matched key words of experiment 2.1.2 [sentences (9)–(12)] vs. experiment 2.2.5 [sentences (22)–(25)]

Sentence	*book*	*table*	*gift*	*mother*
(9)	191	182	170	153
	1.00	1.47	2.07	2.68
(22)	202	194	172	156
	1.00	1.48	3.10	3.70
	man	*truck*	*roses*	*niece*
(10)	187	172	168	161
	1.00	1.43	1.99	2.67
(23)	189	188	169	161
	1.00	1.39	2.96	3.67
	students	*bus*	*gum*	*teacher*
(11)	212	166	165	163
	1.00	1.48	2.03	2.48
(24)	211	179	167	169
	1.00	1.50	3.17	3.62
	ducks	*park*	*popcorn*	*peanuts*
(12)	189	184	173	163
	1.00	1.53	2.16	2.81
(25)	191	190	172	168
	1.00	1.51	3.11	3.73
Grand-means (9)–(12)	195	176	169	160
	1.00	1.47	2.06	2.66
(22)–(25)	199	188	170	163
	1.00	1.47	3.08	3.68

length difference between short and long utterances here was not as great as that in Experiment 2.1.1, or (b) speakers ignore parentheticals in computing utterance length for the purpose of programming a *P1* starting value within the main clause.

Group 2

Given the results of Group 1, we needed a clearer picture of the effect of the parenthetical clause on the F_0 Topline. Three sets of sentences were constructed and spoken by ten new subjects (six male and four female).[3] One of the male subjects had participated in Experiment 2.1.2; the other nine had not

[3]We thank Martha Danly for constructing these sentences.

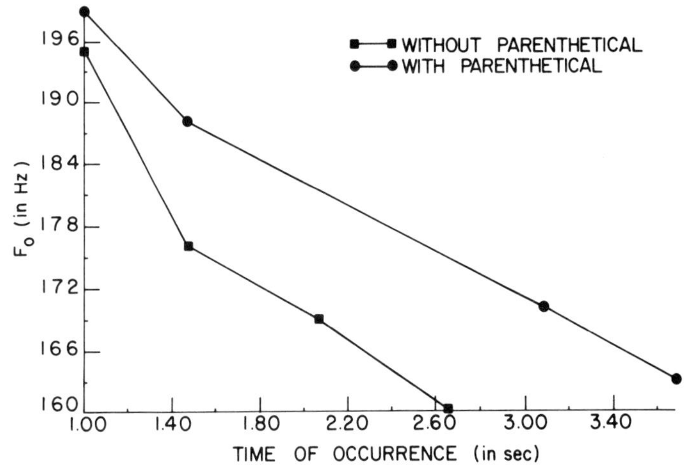

Figure 2.33. Plot of mean F_0 vs. time for each of the measured peak values pooled across sentences without parentheticals from Experiment 2.1.2, and with parentheticals from Experiment 2.2.5.

previously been tested. The sentences of Set 1 contained no parenthetical clauses; those of Set 2 contained a short parenthetical clause *(you realize)*; and those of Set 3 contained a long and more complex parenthetical *(you must have realized by now)*. The sentence sets appear below, with the key words underlined:

Sentence Set 1:

(26) a. *The kids on the bus are waiting for their teacher.*
(27) a. *Some books on the table are ready to be catalogued.*
(28) a. *The bike on the patio belongs to my cousin.*
(29) a. *The house on the corner was built by his grandfather.*

Sentence Set 2:

(26) b. *The kids on the bus, you realize, are waiting for their teacher.*
(27) b. *Some books on the table, you realize, are ready to be catalogued.*
(28) b. *The bike on the patio, you realize, belongs to my cousin.*
(29) b. *The house on the corner, you realize, was built by his grandfather.*

Sentence Set 3:

(26) c. *The kids on the bus, you must have realized by now, are waiting for their teacher.*
(27) c. *Some books on the table, you must have realized by now, are ready to be catalogued.*

(28) c. *The bike on the patio, you must have realized by now, belongs to my cousin.*

(29) c. *The house on the corner, you must have realized by now, was built by his grandfather.*

Because of the complexity of these sentences, commas offsetting the parenthetical clauses appeared on the typed sentence lists exactly as shown here. Each sentence set was included in a separate list during the actual experimental sessions. The order of the sentences differed among each of the three sets.

In each sentence, the four key words were measured for peak F_0 and time of occurrence. The mean F_0 and time of occurrence data for the three sets of sentences are shown in Tables 2.31 through 2.33. The data for individual sentences are plotted in Figures 2.34 through 2.36, and the grand-mean data and Topline Rule are plotted in Figures 2.37 through 2.39. The mean error terms are given in Tables 2.34 through 2.36.

The results of this experiment show that the mean peak F_0 values are approximately equal for the main clause with or without a parenthetical. Sim-

Table 2.31. Means and standard deviations of values of peak F_0 (Hz) and associated time of occurrence (sec) for each of the key words in the set (1) sentences of experiment 2.2.5

Sentence			Measurement			
			P1,T1	*P2,T2*	*P3,T3*	*P4,T4*
(26)	a.	$\bar{F}_0$	183.	155.	146.	149.
		SD	59.6	42.4	45.5	45.8
		$\bar{t}$	1.00	1.47	1.92	2.56
		SD	0.00	0.05	0.14	0.19
(27)	a.	$\bar{F}_0$	173.	165.	142.	145.
		SD	55.9	49.5	41.7	44.4
		$\bar{t}$	1.00	1.52	1.98	2.53
		SD	0.00	0.05	0.11	0.14
(28)	a.	$\bar{F}_0$	171.	161.	145.	146.
		SD	55.5	46.5	41.2	43.0
		$\bar{t}$	1.00	1.45	2.06	2.61
		SD	0.00	0.06	0.12	0.19
(29)	a.	$\bar{F}_0$	176.	168.	152.	145.
		SD	57.7	51.2	45.6	43.2
		$\bar{t}$	1.00	1.48	2.00	2.60
		SD	0.00	0.04	0.11	0.17
Grand-means and SDs		$\bar{F}_0$	176.	162.	146.	146.
		SD	55.1	45.9	42.0	42.4
		$\bar{t}$	1.00	1.48	1.99	2.57
		SD	0.00	0.06	0.13	0.17

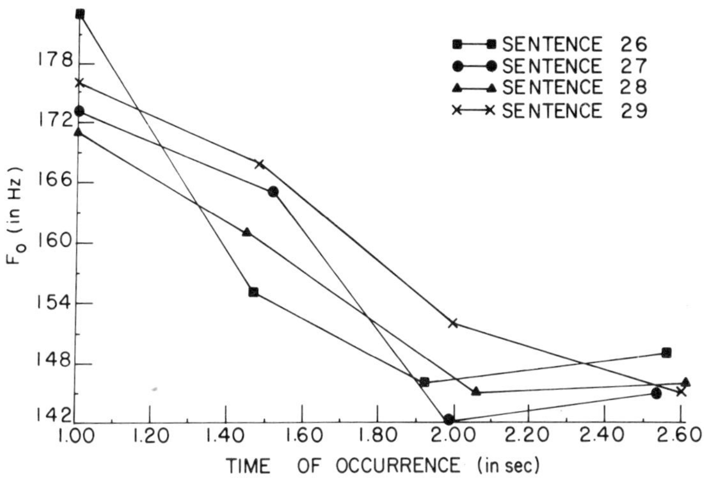

Figure 2.34. Plot of mean F_0 vs. time for each of the measured peak values in Sentences (26a)–(29a).

Table 2.32. Means and standard deviations of values of peak F_0 (Hz) and associated time of occurrence (sec) for each of the key words in the set (2) sentences of experiment 2.2.5

Sentence			Measurement			
			P1,T1	*P2,T2*	*P3,T3*	*P4,T4*
(26)	b.	$\overline{F}_0$	187.	163.	144.	151.
		SD	58.5	44.9	39.5	46.6
		$\bar{t}$	1.00	1.47	2.57	3.17
		SD	0.00	0.06	0.19	0.22
(27)	b.	$\overline{F}_0$	182.	172.	141.	148.
		SD	58.9	52.3	40.8	44.5
		$\bar{t}$	1.00	1.54	2.70	3.24
		SD	0.00	0.08	0.16	0.22
(28)	b.	$\overline{F}_0$	170.	165.	142.	148.
		SD	55.4	41.4	42.7	44.0
		$\bar{t}$	1.00	1.47	2.83	3.37
		SD	0.00	0.06	0.20	0.26
(29)	b.	$\overline{F}_0$	178.	174.	150.	143.
		SD	55.4	47.0	41.2	42.0
		$\bar{t}$	1.00	1.49	2.79	3.38
		SD	0.00	0.04	0.19	0.24
Grand-means and SDs		$\overline{F}_0$	179.	169.	144.	147.
		SD	55.2	45.0	39.6	42.7
		$\bar{t}$	1.00	1.49	2.72	3.29
		SD	0.00	0.06	0.20	0.25

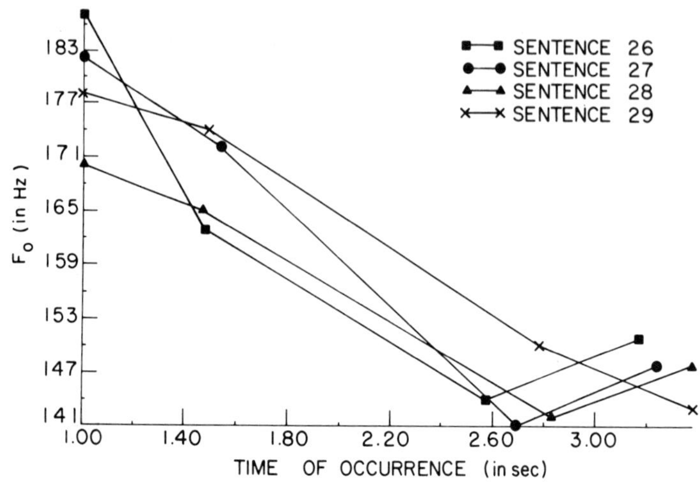

Figure 2.35. Plot of mean F_0 vs. time for each of the measured peak values in Sentences (26b)–(29b).

Table 2.33. Means and standard deviations of values of peak F_0 (Hz) and associated time of occurrence (sec) for each of the key words in the set (3) sentences of experiment 2.2.5

Sentence			Measurement			
			P1,T1	*P2,T2*	*P3,T3*	*P4,T4*
(26)	c.	$\overline{F}_0$	190.	161.	146.	156.
		SD	60.9	47.3	38.6	46.5
		$\bar{t}$	1.00	1.48	3.48	4.10
		SD	0.00	0.07	0.33	0.36
(27)	c.	$\overline{F}_0$	183.	170.	144.	145.
		SD	59.4	52.1	38.5	42.1
		$\bar{t}$	1.00	1.50	3.52	4.05
		SD	0.00	0.05	0.30	0.34
(28)	c.	$\overline{F}_0$	176.	162.	144.	148.
		SD	60.4	44.3	42.3	44.5
		$\bar{t}$	1.00	1.45	3.68	4.19
		SD	0.00	0.07	0.30	0.37
(29)	c.	$\overline{F}_0$	181.	170.	147.	145.
		SD	62.8	53.5	39.5	40.7
		$\bar{t}$	1.00	1.50	3.56	4.15
		SD	0.00	0.06	0.29	0.37
Grand-means and SDs		$\overline{F}_0$	182.	166.	145.	149.
		SD	58.7	47.7	38.2	42.1
		$\bar{t}$	1.00	1.48	3.56	4.12
		SD	0.00	0.06	0.31	0.35

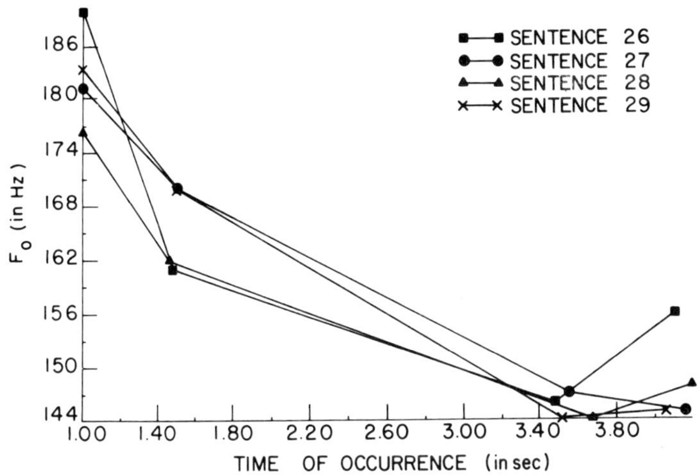

Figure 2.36. Plot of mean F_0 vs. time for each of the measured peak values in Sentences (26c)–(29c).

ilarly, the peak values are about the same for the main clause with short vs. long parentheticals. Individual t tests applied to the peak values in each pairwise comparison of sentences showed only two significant differences out of twelve tests, and these were each observed in a different pairwise sentence comparison, involving a different peak F_0 value.

The major differences between the three sets of sentences involved the slopes of the F_0 topline function, with shallower slopes for the two sentence types containing parentheticals than for the sentences without parentheticals. The slopes for sentences containing longer parentheticals were also shallower than

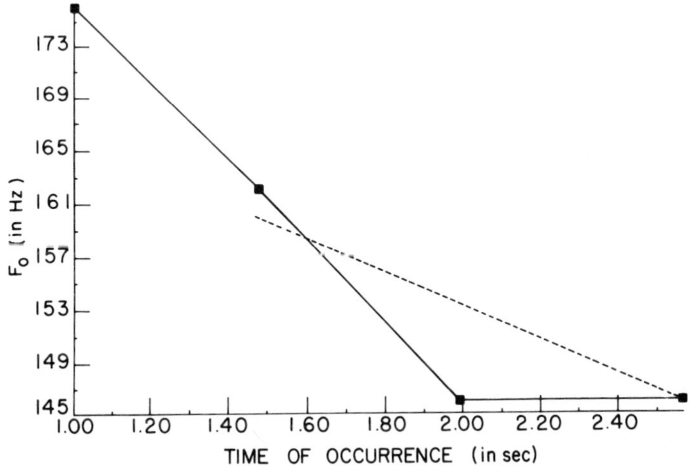

Figure 2.37. Predicted (dotted line) and obtained (solid line) values of mean F_0 vs. time for the grand-mean data of Sentences (26a)–(29a).

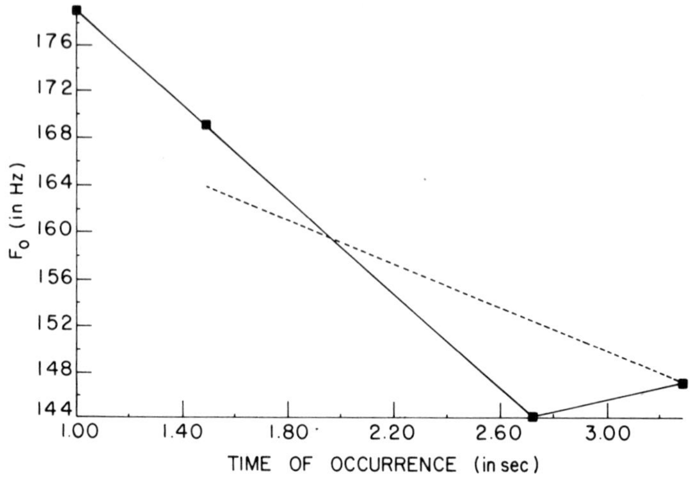

Figure 2.38. Predicted (dotted line) and obtained (solid line) values of mean F_0 vs. time for the grand-mean data of Sentences (26b)–(29b).

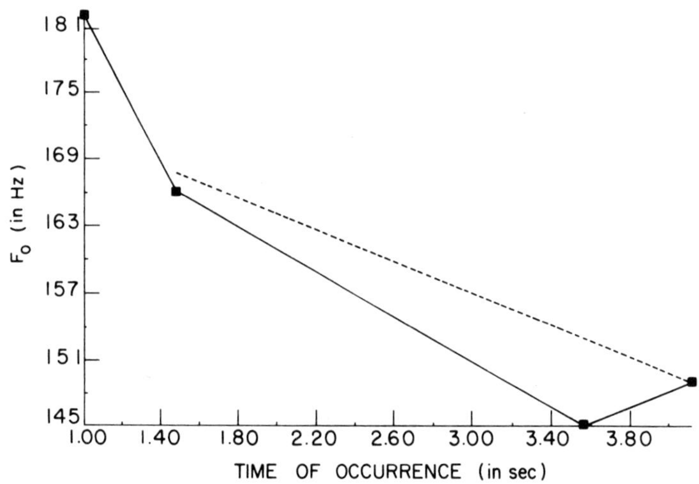

Figure 2.39. Predicted (dotted line) and obtained (solid line) values of mean F_0 vs. time for the grand-mean data of Sentences (26c)–(29c).

the slopes for sentences containing short ones. This difference in slope is a consequence of the fact that the values of the F_0 peaks were relatively fixed across all sentences (see above), while the locations in time for these peaks varied systematically as a function of sentence length. The systematic differences in the slopes are illustrated graphically in Figure 2.40. This figure contains the grand-mean F_0 vs. time of occurrence in the three types of sentences.

The data of Figure 2.40 also show that the introduction of parenthetical material does not alter the temporal relations between key words within the

Table 2.34. Mean and grand-mean percent of error terms of the two predicted versus obtained peak F_0 values in the set (1) sentences of experiment 2.2.5

Sentence			% error in prediction of peak	
			P2	*P3*
(26)	a.	$\overline{E}$	−4.65	−7.46
		SD	9.27	5.45
(27)	a.	$\overline{E}$	5.33	−5.90
		SD	6.22	4.55
(28)	a.	$\overline{E}$	2.47	−4.07
		SD	2.88	1.78
(29)	a	$\overline{E}$	5.52	−0.55
		SD	5.49	3.66
Grand-means and SDs		$\overline{E}$	2.17	−4.49
		SD	7.42	4.72

Table 2.35. Mean and grand-mean percent of error terms of the two predicted versus obtained peak F_0 values in the set (2) sentences of experiment 2.2.5

Sentence			% error in prediction of peak	
			P2	*P3*
(26)	b.	$\overline{E}$	−3.09	−7.44
		SD	5.03	7.42
(27)	b.	$\overline{E}$	4.43	−7.25
		SD	4.54	5.76
(28)	b.	$\overline{E}$	4.98	−5.91
		SD	6.78	4.12
(29)	b.	$\overline{E}$	8.74	1.06
		SD	6.09	3.89
Grand-means and SDs		$\overline{E}$	3.77	−4.89
		SD	6.98	6.34

Table 2.36. Mean and grand-mean percent of error terms of the two predicted vs. obtained peak F_0 values in the set (3) sentences of experiment 2.2.5

Sentence			% error in prediction of peak	
			P2	*P3*
(26)	c.	$\overline{E}$	−7.65	−8.19
		SD	6.19	4.29
(27)	c.	$\overline{E}$	2.14	−3.37
		SD	2.81	4.51
(28)	c.	$\overline{E}$	−0.17	−4.84
		SD	6.37	3.32
(29)	c.	$\overline{E}$	3.17	−0.94
		SD	6.38	4.65
Grand-means and SDs		$\overline{E}$	−0.63	−4.33
		SD	7.01	4.85

main clause that are not separated from one another by the parenthetical material. Note in particular that the time intervals between *P1* and *P2* and between *P3* and *P4* remained relatively fixed in the three types of sentences, with the sole exception of the somewhat longer interval between *P1* and *P2* in Set (1). Thus, the speakers did not speed up or slow down their production of the main clause as a function of the presence of parenthetical material that increased the overall length of the utterance.

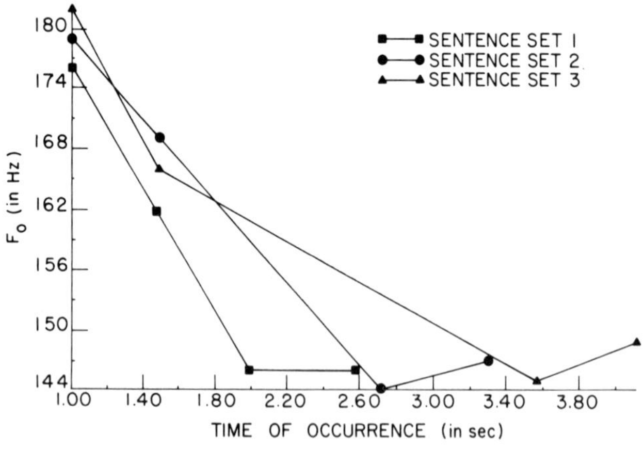

Figure 2.40. Plot of mean F_0 vs. time for each of the measured peak values pooled across sentences of Sets (1)–(3).

The relation between the Topline Rule and the actual topline functions for the sentences appears in Figures 2.37–2.39. The results of these figures show that the Topline Rule more adequately predicts *P2* than *P3* in each of the sentence types, including the base sentences which did NOT include a parenthetical. In all three types of sentences, the mean percent error of the obtained value of *P3* is about −5% (from Tables 2.34–2.36).

To conclude, the results of this experiment indicate that the presence of a parenthetical expression does not seriously perturb the speaker's programming of the main clause topline. The peak values within the main clause were quite similar in sentences with and without parentheticals, suggesting that speakers do not take such parentheticals into account when computing the "length" of the utterance in order to derive a topline function. This finding applies with particular relevance to the first peak in the utterance, since earlier findings (Experiment 2.1.1; also McAllister, 1971; O'Shaughnessy, 1976) showed that higher starting values of this peak are programmed for longer utterances containing no parentheticals. Of additional interest is the finding that speakers could resume their main clause declination function after uttering lower F_0 values during a parenthetical. Further experimentation with sentences containing parentheticals is currently being aimed at addressing the extent to which the speaker is capable of resuming this topline precisely and the possible implications of this phenomenon for models of the speaker's planning and memory of declination.

Study 2.3. The Domain of F_0 Declination

Up to this point, we have examined the F_0 topline in sentences that contain a single main clause. When we extend the investigation to sentences containing more than one main clause, the question arises whether the speaker programs a single topline function over the entire utterance or whether separate functions are programmed for each main clause. By examining the F_0 topline in such sentences, one can determine the domain over which the Topline Rule operates. Previous work has suggested that F_0 declination is indeed "reset" at a major clause boundary (e.g., Lea, 1973; Maeda, 1976), but one must also ask whether such resetting is in fact triggered by the presence of the boundary per se or by an intermediary influence of pausing that is in turn triggered by the presence of the boundary. This latter possibility derives some motivation from the work of Lieberman (1967) on breath groups. According to this possibility, the speaker cannot reset his F_0 declination to a new high starting value without first taking a breath in order to increase the amount of subglottal pressure. Maeda (1976) has suggested, on the other hand, that F_0 resetting could be accomplished without inhalation, and one of the aims of this study was to determine whether resetting could indeed occur in the absence of a pause that is sufficiently long to permit inhalation.

Consider the simple two-clause sentence:

Dick will take the jeep and Clark will drive the truck.

The ideal F_0 topline will take one of two forms for this utterance, depending on whether the "domain of application" of the topline is an entire utterance or a single main clause. If the domain of the F_0 topline covers the entire utterance, then the topline would take the idealized form shown in Figure 2.41(a). The *P1* word, *Dick,* appears above the line given by the Topline Rule and all other words lie on this line. In this case, the Topline Rule is based on (*T1,P1*) and (*T6,P6*), sufficient to model the F_0 topline for the ENTIRE utterance. However, if the domain of application of the F_0 topline is restricted to the syntactically defined main clause, a very different picture emerges. In the case of this example, separate toplines would be applied to each of the two main clauses, and the measured data would ideally be described in Figure 2.41(b). We have

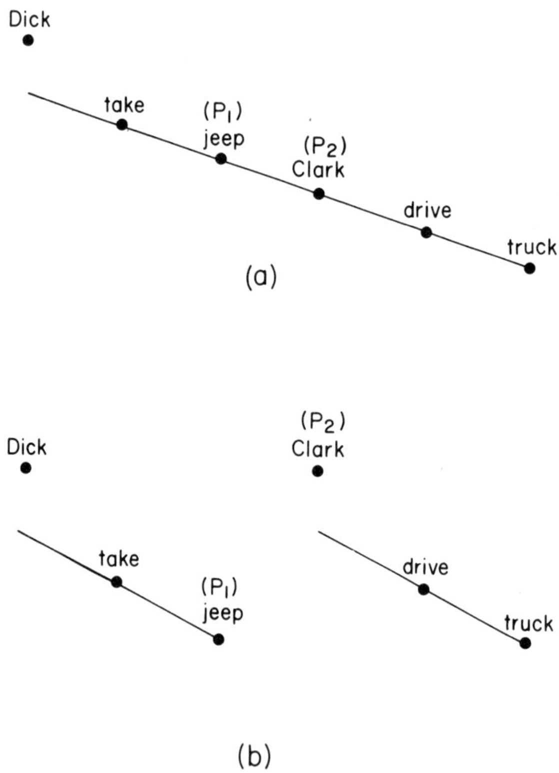

Figure 2.41. Idealization of F_0 topline for different domains of topline application. The top (a) portion shows the idealized topline if the domain of application includes the entire utterance; the bottom (b) portion shows the idealized topline if the domain of application includes a single main clause, given the additional assumption regarding cross-clausal correspondence in F_0 described in the text.

assumed for simplicity that the F_0 values are equal for corresponding words in the two clauses; hence the two lines given by the Topline Rule in the Figure 2.41(b) have identical slope and starting values. A clear implication of Figure 2.41(b) is that all the F_0 values in Clause 2 except the last peak are higher than the last peak F_0 value in Clause 1. In Figure 2.41(a), on the other hand, the F_0 values are monotonically decreasing from the beginning to the end of the utterance. Thus, the relative height of the peak F_0 values in the two clauses should provide a means of determining whether resetting accompanies the clause boundary.

Experiment 2.3.1

Three male and six female subjects participated in this experiment. All had served previously in Experiment 2.1.2. The subjects read the four experimental sentences and two matched fillers following the experimental procedure outlined in Chapter 1. The test sentences, with key words underlined, were as follows:

(30) *The principal of our school warned the pupils to behave but the students on the bus threw gum at the teacher.*

(31) *The chair in the living room was made by my uncle and the book on the table was a gift from my mother.*

(32) *While the men in the truck were eating their lunch the car in the driveway rolled into the street.*

(33) *If the grass near the house is getting too long the boy down the street will mow it tonight.*

Note that the second clauses of Sentences (30) and (31) are identical to Sentences (11) and (9) contained in Experiment 2.1.2. By constructing the sentences in this way, one can compare how the Topline Rule models a clause when it is spoken in isolation as opposed to when it appears as part of a multiclause utterance. In addition, these sentences provide a further test of the Topline Rule by including new variations in the pattern and structure of the key words and clauses.

The mean peak F_0 and time of occurrence for each of the eight key words in each sentence is given in Table 2.37. In Table 2.38, we show a comparison of the second clause data of Sentences (30) and (31) with the data from the same clauses in isolation spoken by the identical nine subjects from Sentences (11) and (9) in Experiment 2.1.2. The data indicate that the peak F_0 was always lower when the clause appeared at the end of the two-clause utterance, and that this difference was greatest for the first peak, generally decreasing in magnitude for each successive peak.

The tabular data of Table 2.37 are also shown graphically in Figure 2.42, which illustrates that the data of Sentence (30) are quite aberrant. Note, for instance, that the mean duration between *T1* and *T2* for Sentence (30) was

Table 2.37. Means and standard deviations of values of peak F_0 (Hz) and associated time of occurrence (sec) for each of the key words in the sentences of experiment 2.3.1

Sentence		Measurement							
		P1,T1	*P2,T2*	*P3,T3*	*P4,T4*	*P5,T5*	*P6,T6*	*P7,T7*	*P8,T8*
(30)	$\bar{F}_0$	207.	185.	187.	172.	195.	164.	162.	167.
	SD	52.4	42.1	43.2	39.8	49.1	43.7	44.6	42.8
	$\bar{t}$	1.00	1.87	2.55	3.16	4.07	4.55	5.05	5.48
	SD	0.00	0.12	0.15	0.19	0.24	0.25	0.29	0.32
(31)	$\bar{F}_0$	210.	194.	174.	166.	181.	175.	168.	156.
	SD	51.4	48.3	44.2	38.4	46.4	39.6	43.0	40.4
	$\bar{t}$	1.00	1.27	2.01	2.49	3.34	3.84	4.43	5.00
	SD	0.00	0.08	0.09	0.13	0.21	0.26	0.27	0.28
(32)	$\bar{F}_0$	197.	189.	181.	161.	178.	173.	161.	164.
	SD	50.1	49.3	47.6	41.8	47.5	45.5	41.4	40.2
	$\bar{t}$	1.00	1.39	1.85	2.23	2.97	3.39	3.95	4.58
	SD	0.00	0.06	0.10	0.13	0.17	0.19	0.21	0.29
(33)	$\bar{F}_0$	203.	180.	172.	164.	177.	170.	160.	156.
	SD	46.6	47.5	47.1	46.5	47.2	44.1	42.9	42.7
	$\bar{t}$	1.00	1.52	1.98	2.46	3.17	3.74	4.13	4.52
	SD	0.00	0.04	0.09	0.11	0.26	0.27	0.29	0.33

870 msec, while the mean duration between *T1* and *T2* of the other three sentences averaged 390 msec. The particular syllabic structure and stress pattern of Sentence (30) can account for this difference. Furthermore, the *P5* value of Sentence (30) was much higher than in the remaining sentences. This high value may be due both to the inherently high F_0 of the vowel /u/ in *students* as well as contrastive stress, typically following the coordinating conjunction *but*. Because the data for Sentence (30) were so different from those of the other sentences, we will not combine them in grouped data to follow.

Table 2.38. Comparison of mean peak F_0 (Hz) in the indicated key word for the identical clauses in experiment 2.1.2 and experiment 2.3.1

Sentence	Key word			
	students	*bus*	*gum*	*teacher*
(11)	213	170	169	169
(30)	195	164	162	167
	book	*table*	*gift*	*mother*
(9)	195	184	173	158
(31)	181	175	168	156

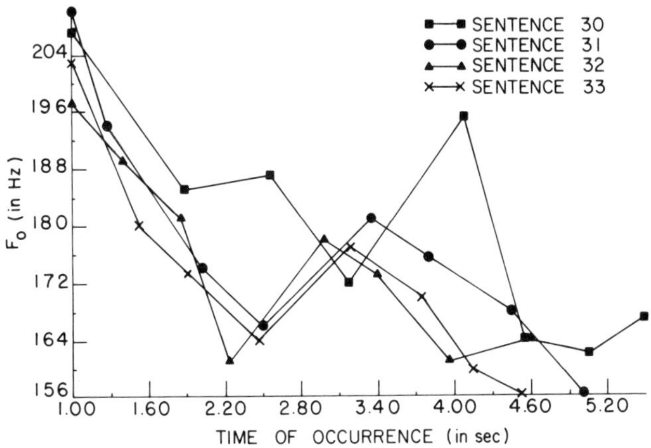

Figure 2.42. Plot of mean F_0 vs. time for each of the measured peak values in Sentences (30)–(33).

Table 2.39 shows the grand-mean data for Sentences (31)–(33). Figure 2.43 shows a plot of these data with the Topline Rule applied to the entire utterance, while Figure 2.44 shows a separate Topline Rule for each clause. In this latter figure, the predicted F_0 falls rather sharply from *P1* to *P4*, rises 15 Hz to *P5* and falls again to *P8*. The actual data fit the predictions of neither Figure 2.43 nor Figure 2.44 very well. In the case of Figure 2.43, one would need to explain the large negative and positive error terms for *P4* and *P5*, respectively. In the case of Figure 2.44, the obtained peak values for two clauses do not have approximately equal slope and starting values.

Since neither of the original hypotheses capture the data, it would seem that a combination of both factors, namely, utterance declination AND clause declination, might provide a better interpretation. In the present experiment, we might suppose that the speaker first plans the F_0 declination of the entire utterance and then the declination of individual clauses, subject to any constraints (range of F_0, value of F_0 peaks at the domain boundaries, etc.) imposed by the overall utterance declination. A great deal more work is needed on this matter, however.

Table 2.39. Grand-means and standard deviations of values of peak F_0 (Hz) and associated time of occurrence (sec) for each of the key words in sentences (31)–(33) of experiment 2.3.1

Grand-means and SDs	Measurement							
	P1,T1	*P2,T2*	*P3,T3*	*P4,T4*	*P5,T5*	*P6,T6*	*P7,T7*	*P8,T8*
$\bar{F}_0$	203.	188.	176.	164.	179.	173.	163.	159.
SD	49.4	48.4	46.3	42.2	47.0	43.1	42.4	41.1
$\bar{t}$	1.00	1.39	1.95	2.39	3.16	3.66	4.17	4.70
SD	0.00	0.06	0.09	0.12	0.21	0.24	0.26	0.30

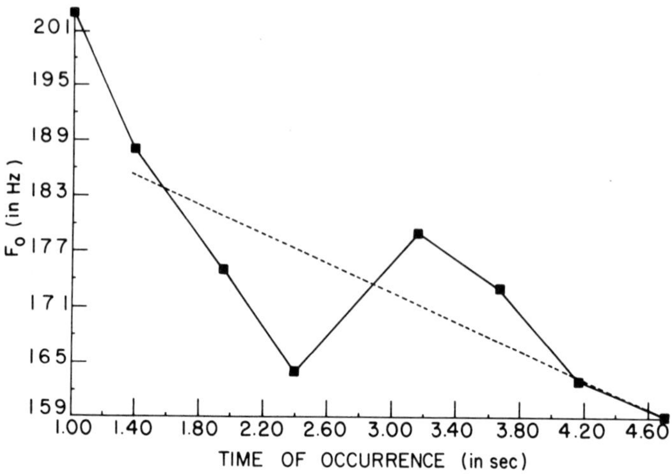

Figure 2.43. Predicted (dotted line) and obtained (solid line) values of mean F_0 vs. time pooled across Sentences (31)–(33). The predicted values were derived under the assumption that the domain of the Topline Rule included the entire utterance.

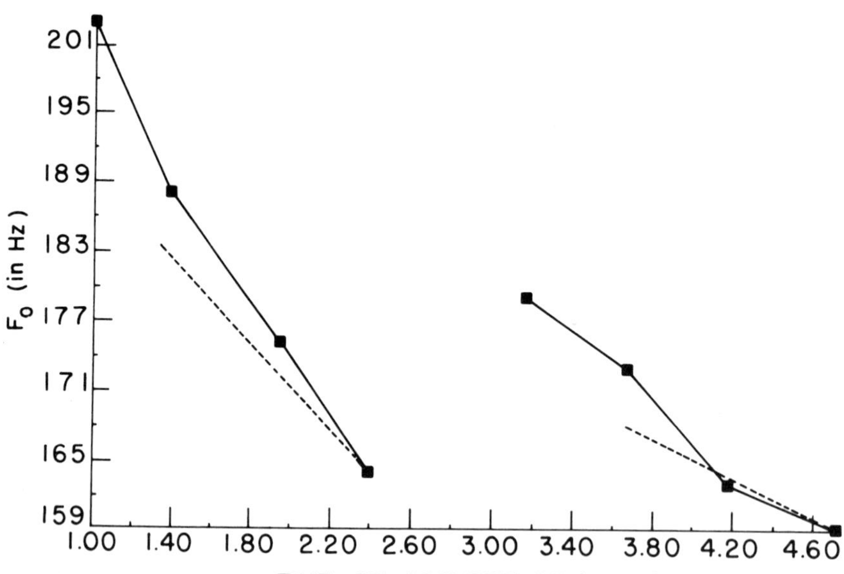

Figure 2.44. Predicted (dotted line) and obtained (solid line) values of mean F_0 vs. time pooled across Sentences (31)–(33). The predicted values were derived under the assumption that the domain of the Topline Rule included a single main clause.

At the outset of this study, we wished to determine if the declination function is "reset" at the beginning of a new clause. Presently, it seems that the speaker does impose some form of declination over the individual clauses in a two-clause utterance, but that, in addition, the entire utterance is subject to a similar process. Therefore, it is correct to say that the declination function is reset at the utterance-internal clause boundary ONLY when referring to that form of declination which the speaker imposes on the individual clauses. The utterance declination is not reset at such a boundary, since by definition this form of declination operates over the domain of the entire utterance. Between utterances, both clausal and utterance types of declination can be said to "reset." In paragraph contexts, it is quite possible that three separate domains of declination are involved, including one each for the domains of clause, sentence, and paragraph. Lea (1973) has observed, for example, that F_0 peaks tend to be successively lower throughout a paragraph (see also Mattingly, 1968; Bolinger, 1970; O'Shaughnessy, 1976), and, relatedly, Lehiste (1975) has reported that the ends of paragraphs may be accompanied by an increment of domain-final segment lengthening.

We shall not attempt here to further refine the Topline Rule to model the various domains of declination. Instead, we shall address the question of what triggers the initiation of a new clause-domain for this attribute.

Experiment 2.3.2

In some cases, the boundary between two main clauses is accompanied by inhalation, and, in such cases, it is impossible to determine whether declination resetting is produced by the presence of the syntactic boundary directly or by the intermediary influence of inhalation, which yields a rise in subglottal pressure. In order to test whether resetting is triggered directly by the syntactic structure, we examined utterances to determine if resetting could occur in the absence of a breath pause.

Ten subjects participated, including six males and four females. One of the male subjects had served in Experiment 2.2.2; the other nine were new to the experimental procedure. The Sentence materials for this experiment consisted of the following two matched pairs:

(34) a. *I plan to bicycle across Canada with <u>Peter</u> <u>Jeanne</u> <u>Fred</u> and <u>Jim</u> sometime this fall.*
b. *I plan to bicycle across Canada with <u>Pete</u> and <u>Jeanne</u> <u>Fred</u> and <u>Jim</u> will drive my car.*

(35) a. *I would like to play basketball with <u>Peter</u> <u>Gene</u> <u>Fred</u> and <u>Jim</u> today after school.*
b. *I would like to play basketball with <u>Pete</u> and <u>Gene</u> <u>Fred</u> and <u>Jim</u> said they would play too.*

The structural distinction between the members of each sentence pair can be schematized by the following bracket notation:

(a) $_S[$*X* <u>*Peter*</u> <u>*Jeanne*</u> <u>*Fred*</u> *and* <u>*Jim*</u> $Y_1]_S$
(b) $_S[$*X* <u>*Pete*</u>$]_S$ $_S[$*and* <u>*Jeanne*</u> <u>*Fred*</u> *and* <u>*Jim*</u> $Y_2]_S$

The speakers were instructed to avoid another possible reading of the (b) sentences, in which, for example, <u>Pete</u> and <u>Jeanne</u> would belong to the first clause and <u>Fred</u> and <u>Jim</u> to the second. The material denoted by Y_1 and Y_2 was matched for overall stress pattern and number of syllables. The bracket notation serves to indicate where the four key words are located in terms of the clause boundaries. The first key word is located in the first main clause in both utterances. The brackets reveal that the second, third, and fourth key words are contained in the first (and only) main clause in (a), while they are found in the second main clause in (b). Based on intuition and the results of Experiment 2.3.1, we hypothesized that the idealized F_0 toplines of the matched sentence pairs appear like those in Figure 2.45. The (b) versions exhibit a "resetting" of the F_0 topline which would provide strong support for the notion that a new domain of declination has begun. Since Figure 2.45 indicates that each key word in Y in the (b) version is higher than its corresponding value in the (a) version, it is unlikely that the difference expected in (b) would be due to local stress heightening.

The peak value of F_0 was measured for each of the four key words in each sentence. The mean peak F_0 for each key word is given in Table 2.40. Table

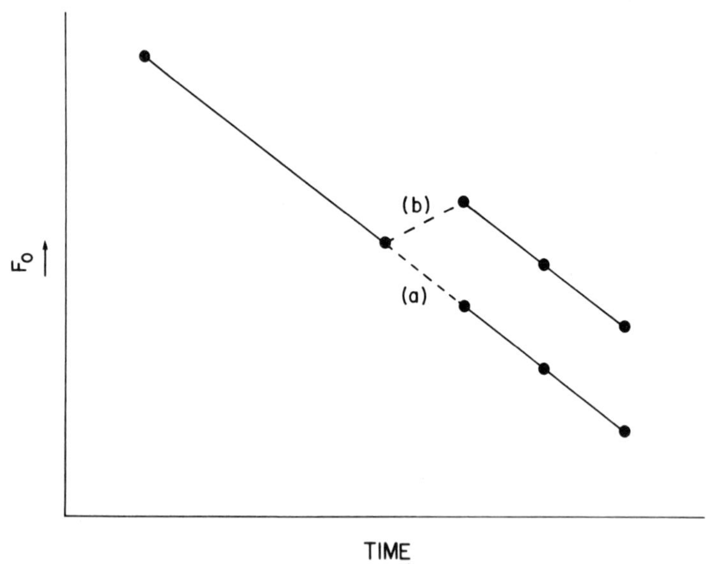

Figure 2.45. Idealized F_0 toplines for the (a) and (b) versions of Sentences (34)–(35). Dotted lines denote the possibility that these intervals represent pauses.

Table 2.40. Means and standard deviations of values of peak F_0 (Hz) for each of the key words in the sentences of experiment 2.3.2

Sentence			Measurement			
			P1	*P2*	*P3*	*P4*
(34)	a.	$\overline{F}_0$	164.	152.	148.	148.
		SD	48.7	46.0	49.4	47.1
	b.	$\overline{F}$	163.	165.	154.	150.
		SD	49.2	45.6	46.6	46.5
(35)	a.	$\overline{F}_0$	165.	151.	148.	150.
		SD	45.7	42.2	44.6	42.3
	b.	$\overline{F}_0$	168.	163.	155.	153.
		SD	45.1	52.6	53.0	47.5

2.41 shows the results of a series of two-tailed *t* tests for matched pairs, comparing the peak F_0 value of each corresponding key word within each sentence pair.

Considering the key word pairs one at a time, let us first examine *P1*. In both sentence pairs, this test pitted the peak F_0 of *Peter* vs. *Pete,* and in both cases there was no significant difference. The table of mean peak F_0 (Table 2.40) indicates that, in Pair (34), *Peter* was 1 Hz higher, while in Pair (35), *Pete* was higher by 3 Hz. The data support the notion that the F_0 topline prior to the clause boundary is similar in the two sentences. For the next three key words, which differed in their location in the clause structure between the (a) and (b) sentences (refer back to the structural descriptions), the peak F_0 was higher in the two-clause version in every case. This difference was significant for the key words *Jeanne* in Pair (34) and *Gene* in Pair (35) (representing *P2*); *Fred* in both Pairs (representing *P3*); and nonsignificant for *Jim* in both Pairs (representing *P4*). The effects for *P2* through *P4* suggest that the presence of the clause boundary in the (b) sentences triggered declination resetting.

In conclusion, the presence of a clause boundary in the (b) versions triggered

Table 2.41. *t*-Tests on values of peak F_0 in matched key words for experiment 2.3.2

(b) vs. (a) in sentence pair	Measurement			
	P1	*P2*	*P3*	*P4*
(34)	−0.18	4.51^c	4.23^c	0.72
(35)	1.00	3.14^b	2.39^a	0.87

[a] $p < .05$.
[b] $p < .02$.
[c] $p < .005$.

no change in the peak F_0 value measured immediately prior to the boundary but did yield significant raising of the first two peaks following the boundary. Since the third F_0 peak following the boundary was essentially unchanged, however, we must reject the strong F_0 resetting hypothesis represented in Figure 2.45. In fact, since utterances which differ substantially in features such as length tend to yield approximately equal F_0 values for the final peak (e.g., Experiment 2.1.1), there exist independent grounds for rejecting this strong hypothesis. In view of the data of Experiment 2.3.1, it appears again that a form of declination operates separately over both utterance and clause domains.

Let us now consider whether the new domain of F_0 declination appearing at the clause boundary in the (b) sentences is triggered DIRECTLY by the syntactic boundary or if the boundary triggers a breathing pause, which in turn initiates the new domain of declination. Pause durations were measured using the AUDITS computer program (Huggins, 1969) from the cessation of voicing for *Pete* to the onset of voicing for *and* in Sentences (34b) and (35b). AUDITS allows the experimenter to measure segmental durations by visual inspection of an oscillographic trace. The pause data exhibited the following means and standard deviations:

Sentence (34b): $\overline{X}$ = 337.0 msec, SD = 197.6 msec
Sentence (35b): $\overline{X}$ = 303.8 msec, SD = 172.4 msec

The large standard deviations indicate a considerable range of pause durations across the ten subjects. In Sentence (34b), the pause durations ranged from 0 to 739.4 msec, and in (35b), from 0 to 549.9 msec. The first conclusion to be drawn from these data is that pausing is not obligatory at the site between two conjoined clauses. Since the speaker must pause in order to take a breath, we can also state that breathing is not obligatory at this location. Stevens (personal communication) has suggested that it is unlikely that speakers could execute an inspiration in less than 150 msec (see also Grosjean and Collins, 1979). In the 20 sentences measured for pauses above, five contained pauses of 192 msec or less with a mean pause duration of only 84 msec. It seems unlikely that breaths would have occurred in these sentences. Furthermore, the high signal to noise ratio of the room in which the recordings were made allowed us to actually hear the inspiration of a subject. Only four of the 20 utterances were noted to have perceptible inspirations. The mean duration of these four pauses was 428.5 msec, with an SD of 63.4 msec.

Of the five utterances containing pauses of less than 200 msec, four showed resetting. We chose the pattern of results found in Table 2.40 as a criterion for resetting in a single (b) vs. (a) sentence pair, comparing the individual F_0 values of the five (b) sentences containing short pauses with those of the corresponding (a) sentence. If a given sentence pair showed approximately the same pattern of results as those obtained in Table 2.40, it was concluded that resetting had occurred. Over all four pairs, the average values for peak F_0 (*P1* through *P4*) for the (b) versions were 173, 171, 160, and 164 Hz, and for the

(a) versions, 166, 153, 148, and 150 Hz. Two of the speakers were male, two were female, and both sentence pairs were represented twice in the four pairs. Thus, no single speaker or sentence pair was responsible for producing resetting in the absence of a breathing pause. The differences between paired values of *P2*, *P3*, and *P4* averaged 18, 12, and 14 Hz, respectively. These differences actually show a greater magnitude of change in (b) vs. (a) than the pooled data. The data indicate that the new domain of F_0 declination at a clause boundary is not dependent on the presence of a pause. Furthermore, one sentence pair for one of the two subjects with a perceptible breathing pause in the (b) version showed no resetting at all. We can thus tentatively conclude that the conjoined clause boundary optionally triggers both resetting and breathing, and that, while resetting and breathing often co-occur, each is not responsible for the presence of the other. Ultimately, measurements of breathing will need to be obtained for sentences such as those above, but our acoustic and perceptual evidence strongly suggest that resetting and breathing are independent, both probabilistically triggered by the presence of a major syntactic boundary and other factors.

In the absence of a breathing pause to increase subglottal air pressure by refilling the lungs, how does a speaker physiologically accomplish F_0 resetting in a case such as this? Maeda (1976) has suggested that an increase in the activity of the abdominal muscles could push the diaphragm upward and thereby increase the subglottal pressure by an amount sufficient to allow resetting without inhalation, consistent with our data.

General Conclusions

The studies of this chapter indicate that the form of declination can be captured by an abstract mathematical formula—the Topline Rule—which adequately predicts the values of intermediate peak F_0 throughout the course of a large variety of single-clause declarative sentence structures in English. This Topline Rule is approximately invariant across typical variations in the following characteristics:

(1) Sex of speaker
(2) Length of clause
(3) Type of clause
(4) Type of grammatical category, excluding Adverbs
(5) Number of measured peaks per clause
(6) Unemphatic stress patterns
(7) Rate of speech
(8) Presence or absence of intervening parenthetical clause

The fact that the Topline Rule adequately predicts peak F_0 in utterances varying in these diverse characteristics suggests that the rule has indeed captured a good deal of the essence of the declination attribute. In many cases, the mean percent of error in peak value predictions averaged less than 2%, well within the range of accuracy required for applications such as speech synthesis (see Chapter 6).

While the Topline Rule provides a good characterization of the declination of peaks in declarative sentences of English, it is equally clear that this formulation does not constitute a linguistic universal. For example, in collaboration with Kazuhiko Yorifuji, we have recently completed a study of declination in Japanese, and the results point to major differences between the form of declination in Japanese and English (Yorifuji, Cooper, and Sorensen, 1978).

As a pitch accent language, Japanese includes major differences in the inherent F_0 of individual syllables, unlike English (McCawley, 1968). For this reason, one would expect that Japanese would not adhere whatsoever to the Topline Rule in individual utterances, although the Topline Rule might be manifest as a "dc" level declination when averaged across a large variety of utterances. As a verb-final language, Japanese also differs radically from English in its basic syntactic structure, with the ordering of major constituents typically being Subject–Object–Verb in Japanese and Subject–Verb–Object in English (Kuno, 1973).

In our Japanese study, 6 native speakers of the standard Tokyo dialect of Japanese were asked to read lists of Japanese sentences that were printed using kanji and hiragana characters. The instructions were presented to the speakers in Japanese following the procedural details of our studies of English. Twelve test sentences were included in the main experiment, consisting of six pairs of matched short and long sentences each containing three key words, following the design of Experiment 2.1.1 for English. The results showed two general phenomena that were also found in the study of English, namely, (1) the presence of some form of declination for both peaks and valleys throughout an utterance, and (2) a steeper slope for the declination of the peaks than the declination of the valleys. Both these results are consistent with our English data. However, the precise form of the declination for the peaks in Japanese did not conform well to the Topline Rule generated for English. This finding is hardly surprising, since Japanese exhibits lexical pitch accent on individual words and generally seems to contain shorter domains of F_0 programming than in English (see also Fujisaki and Omura, 1971, for related durational characteristics). This difference between the two languages was also manifest in the values of the first peak of the utterances. Whereas in English *P1* exhibits a higher value in long vs. short sentences (Experiment 2.2.1; McAllister, 1971; O'Shaughnessy, 1976), the values of *P1* in Japanese were approximately equal in these two versions, suggesting that Japanese speakers do not take the length of the entire sentence into consideration when programming their first F_0 peak. Whereas the values of *P1* in long vs. short sentences differed for English but not Japanese, the values of *P3* (and *V3*) in these same sentences differed for Japanese but not English. Recall that, in English, the values of *P3* were approximately equal in long vs. short sentences. In Japanese, the values of both *P3* and *V3* were consistently higher in short sentences, suggesting that declination in Japanese proceeds at a more nearly constant rate for long vs. short utterances than in English.

In sum, a comparison of English and Japanese reveals that both languages exhibit the general feature of declination but that the Topline Rule is applicable only to English. Japanese speakers appear to program their F_0 values over shorter domains than in English, and the inherent pitch accent of individual words plays a much greater role in the F_0 values for Japanese. In effect, the local effects of lexical accent outweigh any systematic global effects of declination to such an extent that a declination function cannot be readily ascertained, and, while further work might eventually uncover such a function, it is unlikely that it will bear close resemblance to the Topline Rule for English.

Although the aim of this chapter has been primarily to model the form of English declination quantitatively as a prerequisite to studying local F_0 attributes typically superimposed on declination, these studies have also led to some inferences about the psychological processes that underly the production of F_0. For example, evidence for long-range planning operations was provided by examining the peak values at the beginning of long vs. short sentences. The higher F_0 values obtained for the first peak of a longer sentence indicated that English speakers program this value by taking into account a general consideration of the overall length/complexity of the sentence.

The Topline Rule also provided a useful landmark for discussing cases in which this rule did *not* adequately predict peak F_0 values. For example, in the experiment on speaking rates, the obtained peak values in fast sentences were consistently higher than those predicted from the Topline Rule, whereas the obtained peaks in slow sentences were consistently lower. The directions of these two displacements are accountable in terms of vocal fold tension as a general speaking posture in fast vs. slow speech. Without the "dc" component provided by the Topline Rule, however, it would be more difficult to discern the presence and magnitude of this additional factor. The same point can be made about the application of the Topline Rule to key words bearing different inherent vowels. Again, the direction of displacement between the predicted and obtained values can often be predicted on the basis of the combined factors of the Topline Rule and inherent F_0 of the particular vowel.

Finally, the results of this chapter suggest that the declination function may be reset at the boundary between two main clauses of at least moderate length. Additional evidence indicates that entire utterances exhibit a domain of declination of their own. Resetting seems to be triggered directly by the presence of the syntactic boundary and does not depend on an accompanying breath pause. While resetting has been examined here from the standpoint of syntactic structure, it is also suspected that a variety of nonsyntactic factors exert some influence on the likelihood of resetting at a given clause boundary. In particular, resetting should be more likely in the environment of longer clauses, slower rates of speech, and lower semantic relatedness between the two clauses. These factors will be discussed further in Chapter 5.

We have focused our attention in this chapter on the problem of predicting F_0 peaks, with less concern about predicting other global attributes of F_0 declination. While there can be no question that much salient information is contained in the peak values, it seems equally clear that an adequate theory of declination must include a characterization of non-peak values as well.

Taking a first step to develop such a theory, let us ask whether the peak values can be useful in predicting non-peak values. Intuitively, the general answer to this question seems "yes," under the assumption that speakers actively program target peaks in F_0, with non-peak portions of the F_0 contour determined passively by physiological constraints on such relevant factors as subglottal pressure and vocal fold tension. In some cases, non-peak values are certainly under active programming, as in the case of continuation rises (see Chapter 1), but, as a first approximation, it seems that peaks in F_0 represent the most actively programmed attribute. If so, then the peaks should certainly be useful in predicting other aspects of the F_0 contour.

Recently, Pierrehumbert (1979a) has adopted this viewpoint in attempting to construct a model of F_0 contours for speech synthesis. She has expressed the idea that the peak values act as anchor points and that the non-peak values can be determined in a manner that partially mimics the draping of a loose string through the peaks, as shown in Figure 2.46. The anchor-string analogy has stimulated us to ask a number of related questions.

First, some clarification of the anchor-string model is in order. The peak values themselves seem to act more like pulleys than fixed anchor points, since the string tension is roughly constant regardless of the distance between adjacent peaks. In addition, the string model must be modified to correctly predict thc form of thc F_0 pattern in the immediate region of the peak values. According to the present model, the form of this pattern is primarily concave upwards, whereas actual patterns typically exhibit a concave downward form (see Figure 2.47). In colloquial terms, the region of the peak more nearly resembles the

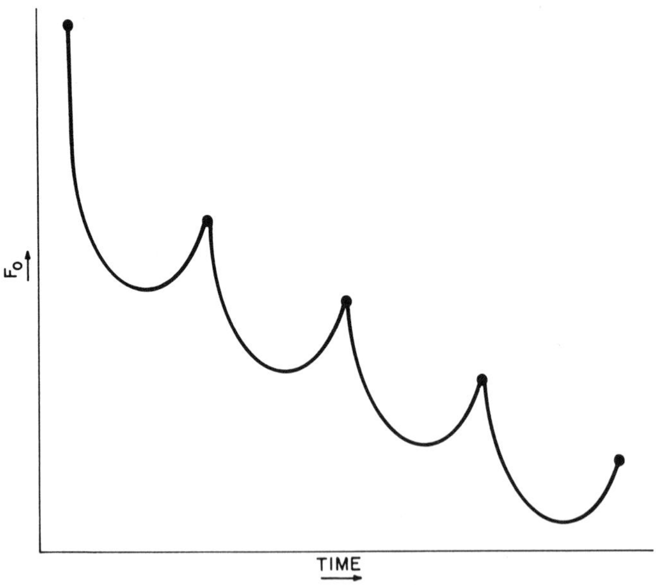

Figure 2.46. String model for predicting nonpeak values of F_0 in a sentence given only the peak values, following a suggestion of Pierrehumbert (1979).

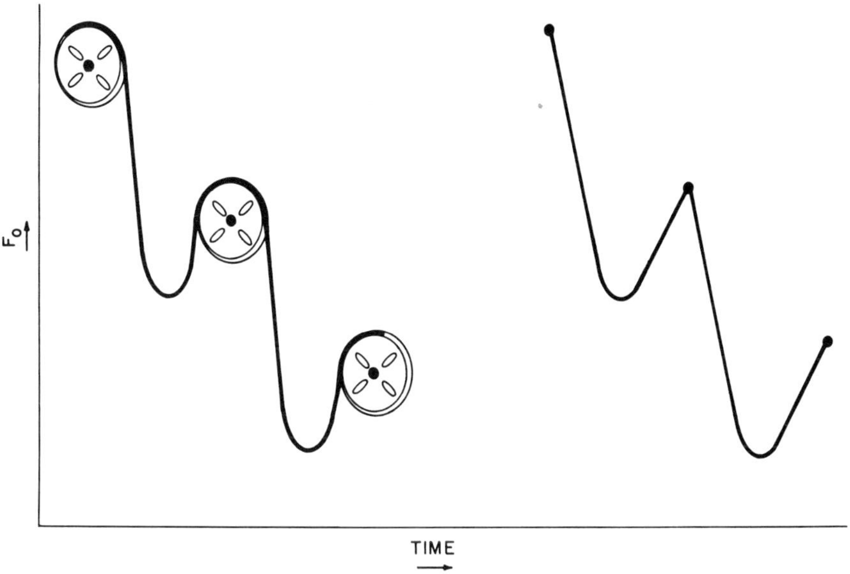

Figure 2.47. Revised string model in which the diameters of the pulleys are enlarged (left) in order to capture the concave downward shape of F_0 patterns in the region near peak values (see text for description).

Appalachians than the Alps. In order to capture this pattern, the pulleys themselves are conceived as being relatively large in diameter, as shown in Figure 2.47.

Given these modifications, the pulley-string model makes at least one interesting prediction about the location of valley values. In particular, the model predicts that the valley should be located more near the peak having a lower F_0 value of the two peaks surrounding the valley. As shown in Figure 2.48, the valley should be displaced toward *P2* to the extent that *P1* is higher in F_0 than

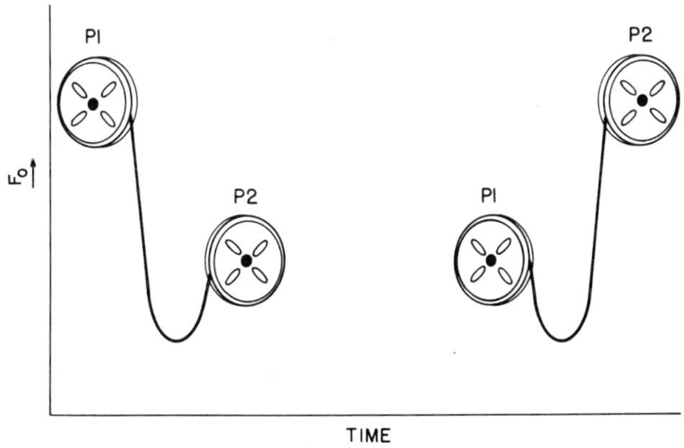

Figure 2.48. String model showing that the predicted location of a valley between two peaks differs depending on the relative heights of the peaks.

P2, and the valley should be displaced toward *P1* to the extent that *P2* is higher than *P1*. The data of Olive (1975), O'Shaughnessy (1976), and our own informal observations support this general prediction, although it remains to be determined whether the precise location of the valley varies in direct proportion to the relative heights of *P1* and *P2*.

The pulley-string model suggests that values of peak F_0 may indeed be useful in predicting the locations and values of other aspects of the F_0 contour. If so, it would be possible to predict most of the F_0 pattern of a sentence from very few predetermined values, including the first and last peaks of the sentence, in order to compute the Topline Rule, at least one valley, to determine the string tension, and any local effects of stress pattern.[4] Such a model would be far more predictive than a complex model such as Olive's (1975), in which parameters must be determined for each individual word in the sentence. The current Topline Rule predicts the values of sentence-internal peaks in Olive's corpus with about the same degree of accuracy as his complex model.

Finally, it should be noted that any attempts at abstract modeling for peaks and valleys must be aimed at capturing an average F_0 contour, such that a variety of phonetic effects and other influences on F_0 can be superimposed onto such a contour to derive predictions for an actual utterance. This strategy stems from the assumption that an average contour does indeed represent the essence of F_0 declination and that this declination forms an independent portion of the overall F_0 contour. There is, however, no assurance that all local influences on F_0 can be superimposed on declination in a simple way.

[4]Note that the model will not handle utterance-final fall nor internal continuation rises. The string tension seems to slacken somewhat from the beginning to the end of an utterance, and Olive (1975) has incorporated this feature into his polynomial model of F_0 patterns. Also, shorter words would seem to have flatter F_0 contours in general. Olive proposes a special rule for the determiner *the*, and this rule or some form of it probably also applies to many other words.

3. Fall–Rise Patterns

As noted in Chapter 1, local fall–rise patterns of F_0 typically accompany the boundaries of clauses and phrases and are operationally distinguishable from declination resetting by virtue of their different domains. Fall–rise patterns are typically superimposed over no more than two words, whereas declination resetting is manifest throughout a longer stretch of speech. In general, fall–rise patterns accompany major syntactic boundaries whether or not these boundaries are also accompanied by a new domain of F_0 declination. A detailed examination of such patterns is undertaken here in an attempt to provide information about some of the units of syntactic structure computed by the speaker. We shall consider fall–rise patterns at major clause boundaries as well as at phrase boundaries within a clause.

In each study to be presented, the falling portion of the fall–rise pattern is contained in a single stressed syllable and is measured from peak (*P1*) to valley (*V*). The rising portion of the pattern is measured from the valley of this same syllable to the peak (*P2*) at the beginning of the next stressed word. In effect, then, the F_0 fall is measured within the confines of a single syllable, whereas the F_0 rise is measured across a word boundary.

The most general issue raised in these studies involves the relative magnitude of fall–rise patterns at various syntactic boundaries. According to the theory of syntactic boundary strengths proposed by Cooper and Paccia-Cooper (1980), stronger syntactic boundaries are typically accompanied by more segmental lengthening, pausing, blocking of phonological rules, and fall–rise patterns in F_0. In the earlier work on speech timing, a fairly detailed metric of boundary strengths was developed to provide an account of relative differences in the magnitudes of segmental lengthening and pausing at a number of different syntactic boundaries, but the metric has yet to be tested for its ability to predict the rank magnitude of fall–rise patterns in F_0. Previous studies of fall–rise patterns by Lea (1972, 1973) and O'Shaughnessy (1976) already suggest that the rank magnitude of such patterns is determined, at least in part, by the strength of the intervening syntactic boundary. With the explicit strength metric presented below, combined with experimental testing of controlled sentence materials, it is now possible to test both broadbrush and fine-grained aspects of the boundary strength hypothesis.

According to this metric, the strength of a given boundary is determined primarily by the BRANCHING DEPTH of the nodes which dominate the words immediately preceding and following the boundary, with additional weight assigned to the dominating node preceding the boundary and Sentence nodes. An algorithm for determining the syntactic boundary strength can be specified as follows (Cooper and Paccia-Cooper, 1980):

Step 1. Define the structural representation of the utterance, using rewrite rules of phrase structure grammar.

Step 2. Locate the key boundary.

Step 3. Locate the dominating phrase structure nodes by finding the highest nodes which dominate the word to the immediate left or right of the key boundary but which do not dominate both of these words.

Step 4. Determine the number of flanking nodes between the dominating phrase structure nodes and the nodes immediately dominating the words to the immediate left and right of the key boundary, assigning a value of 0 to (a) any nodes referring to minor category terms (including conjunctions, determiners, and non-lexical prepositions) and (b) when considering the left side of the boundary, any nonterminal nodes that do not branch. Such nodes include those that do not branch into at least two nodes, each of which dominates a major category term.

Step 5. Add one unit of strength to any branching S node.

Step 6. Multiply by two the number of nodes listed for the left side of the boundary.

Step 7. Combine the number of nodes for both left and right sides of the boundary.

The output of this metric is an integer of strength from 0 to n, typically ranging from 1 (weakest) to about 7 (strongest). This information is then fed into an algorithm for determining the probability and magnitude of a prosodic characteristic, considering additional factors such as speaking rate and constituent length. Additional refinements of the metric have been suggested, but the basic format shown here will serve adequately for the issues to be discussed throughout this chapter.

Fall–rise patterns offer a particularly good opportunity to evaluate the most fundamental aspects of this metric (Steps 1–4), since the individual falling and rising components of such patterns may be differentially sensitive to the weightings of phrase nodes that immediately precede vs. follow the syntactic boundary under consideration. In previous work, differences in the contributions of preceding and following phrase nodes were observed for segmental lengthening and the blocking of cross-word phonetic conditioning; but, because fall-rise patterns contain two separate components, the individual contributions of each bordering phrase can now be studied better within the same utterances.

Study 3.1. Clause Boundaries

This study contains three experiments in which fall–rise patterns were examined at clause boundaries. In each case, an attempt was made to compare the magnitudes of fall–rise patterns for matched phonetic material in the region of different syntactic boundaries, occurring at comparable locations in distinct sentences. With this experimental design, it was possible to provide various tests of the general hypothesis that stronger syntactic boundaries are accompanied by larger fall–rise patterns than weaker boundaries.

Experiment 3.1.1

In this experiment, we studied fall–rise patterns occurring across two key words that either spanned a major clause boundary or appeared within the same conjoined Noun Phrase (NP). We hypothesized that fall–rise patterns would be more pronounced in the former region, based on the proposed metric of syntactic boundary strengths. This prediction constitutes a broadbrush claim that would be incorporated in virtually any version of a boundary strength metric.

Seven males and three females participated in this experiment. The sentence materials included two matched pairs:

(1) a. *I decided to go to the party this weekend with Steve and Tina is planning to go with Tim.*
    b. *I decided to go to the party this weekend with Steve and Tina if Patty will go with Tim.*
(2) a. *I decided to go to the party this weekend with Tina and Steve is planning to go with Lynn.*
    b. *I decided to go to the party this weekend with Tina and Steve if Patty will go with Tim.*

The basic clause structure for the members of each sentence pair appears below, using Pair (1) as an example:

(1′) a. $_S[X\ \underline{Steve}]_S\ _S[and\ \underline{Tina}\ Y_1]_S$
     b. $_S[X\ \underline{Steve}\ and\ \underline{Tina}]_S\ _S[Y_2]_S$

In both sentences, note that the phonetic material is identical from the beginning of the sentence to the end of the key word *Tina*. The following material, denoted by Y_1 and Y_2, is matched as closely as possible for phonetic content, stress pattern, and number of syllables.

As shown by the bracketing of (1′a), a substantial syntactic boundary exists between the two underlined key words, whereas, in (b), no large syntactic

Table 3.1. Mean F_0 values for experiment 3.1.1

Sentence		*P1*	*V*	*P2*
(1)	a.	155	127	171
	b.	153	147	158
(2)	a.	161	144	163
	b.	162	154	152

boundary intervenes between the same key words. Hence, we expect a greater fall–rise pattern in (a) vs. (b).

The sentence pairs were presented to the subjects with six matched filler sentences. After recording, an F_0 contour of the key words was obtained using the FPRD program. Three F_0 measurements were recorded as data for each utterance. These included the peak (*P1*) and valley (*V*) value of F_0 in the first key word and the peak (*P2*) value in the second key word. When the bisyllabic word *Tina* appeared as the first key word, both *P1* and *V* were measured within the stressed syllable. The means of these measurements are given in Table 3.1. The data of mean F_0 at each of the three locations are shown in Figure 3.1 [Sentence Pair (1)] and Figure 3.2 [Sentence Pair (2)].

P1 was significantly greater than *V* in all 4 individual sentences ($p < .001$, t values ranging from 5.87 to 7.25, $df = 9$). As predicted, the magnitude of this fall, however, was quite different for the (a) vs. (b) sentences, as seen in Figures 3.1 and 3.2. The amount of fall (*P1*–*V*) in individual sentences was significantly greater for both (a) versions: [(1a) vs. (1b), $p < .02$, $t = 3.07$; and (2a) vs. (2b), $p < .05$, $t = 2.45$, $df = 9$ for both tests]. As shown in

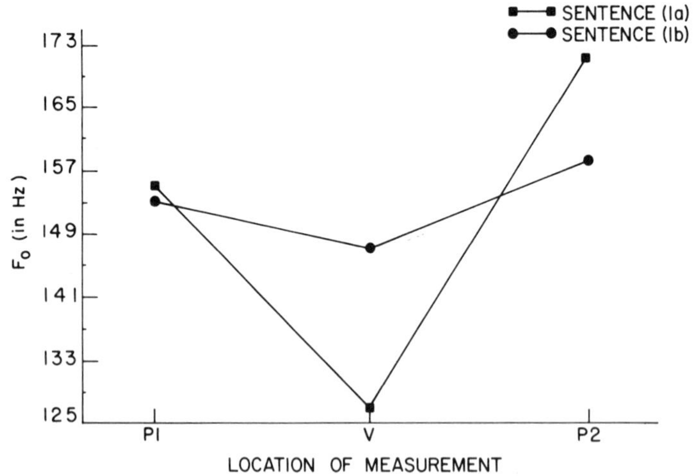

Figure 3.1. Plot of mean F_0 vs. time for each of the measured values in Sentence Pair (1).

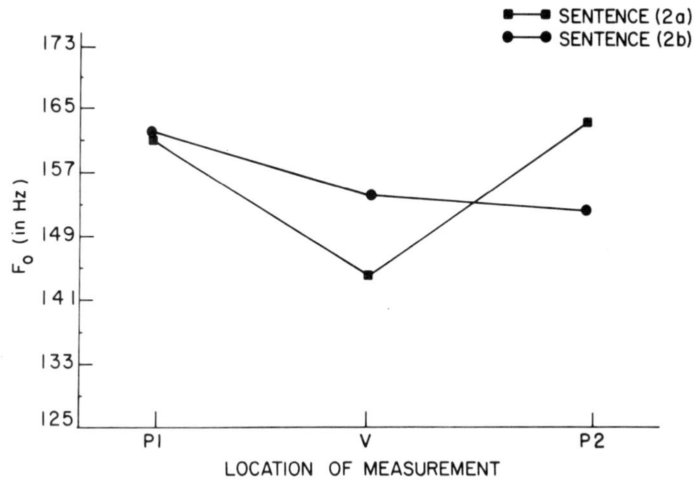

Figure 3.2. Plot of mean F_0 vs. time for each of the measured values in Sentence Pair (2).

Figures 3.1 and 3.2, the significantly greater fall in (a) is attributable to a lower *V* value in both (a) versions, with *P1* values being nearly identical in the (a) vs. (b) sentences. The speaker thus signaled the end of a clause by producing a substantial F_0 fall without raising the F_0 peak at the beginning of the clause-final word. This result suggests that the F_0 fall accompanying this location is not explicable in terms of a local effect of stress heightening for the clause-final word.

Considering again the data within individual sentences, the rise from *V* to *P2* was statistically significant in Sentences (1a), (1b), and (2a) ($p < .01$, t values ranging from 3.48 to 6.75, $df = 9$). A nonsignificant trend in the opposite direction was obtained in Sentence (2b) ($p > .10$, $t = 1.61$, $df = 9$). Comparing the amount of rise (*V–P2*) in the (a) vs. (b) sentences, the rise was greater in both (a) sentences ($p < .01$, t values of 3.70 and 6.70, $df = 9$). As can be seen in Figures 3.1 and 3.2, the significantly greater rise in F_0 from *V* to *P2* in the (a) versions is attributable to BOTH the lower values of *V* and the higher values of *P2* in these versions. The higher values of *P2* in the (a) versions may be explained in relation to the fact that the *P2* word begins a new main clause. As observed in Study 2.4, the beginning of such a clause may be accompanied by resetting of the F_0 declination function, accounting for the higher value for *P2* in the (a) versions here.

While significant fall–rise patterns were obtained for the (a) versions of both Sentences (1) and (2), a comparison of Figures 3.1 and 3.2 reveals that the magnitude of both the falling and rising portion of the F_0 pattern was about twice as large in (1a) as in (2a). This difference in magnitude may be primarily attributable to the phonetic influence of the different key words preceding the boundary, *Steve* in (1a) and *Tina* in (2a). Recall that both *P1* and *V* were measured for the stressed syllable/ti/ in *Tina.* The vowel /i/ is typically shorter when it appears in a trochaic bisyllable like *Tina* (e.g., Huggins, 1974; Cooper, Lapointe, and Paccia, 1977) than when it appears in a monosyllabic word like *Steve,* and it is suggested that the greater fall in F_0 observed in (1a) may be attributable to the greater amount of time afforded the speaker to execute this pattern. While data to be reviewed in Chapter 5 indicate that the magnitude of an F_0 fall is not necessarily correlated with segmental duration, a certain minimum duration is required in order for a speaker to execute a large F_0 fall. The greater rise in F_0 may be similarly explained, since the difference in the magnitude of the rise in (1a) vs. (2a) was produced largely by the much lower average value of V in (1a), measured for the first key word *Steve.*

Returning to compare the values of *P2* in the (a) vs. (b) versions of both sentences, let us recall that the higher values of *P2* in the (a) versions may be attributable to the fact that the *P2* word in (a) begins a new main clause. There exists an alternative possibility, however, relying on a difference in grammatical relations. Perhaps, according to this view, *P2* was higher in the (a) versions because the *P2* word serves as a subject of the clause in these versions, whereas the *P2* word functions as part of a conjoined direct object in the (b) versions. To test this possibility, we conducted an experiment to assess any inherent differences in the F_0 peaks of words serving as subject vs. object. The experiment included two test sentences and eight filler sentences, read by ten speakers. The test sentences appear below:

(3) a. *Jim and his best friend Pete saw Sue while Bob saw Jeanne and her roommate Joanne.*
 b. *Jill and her roommate Jeanne saw Pete while Sue saw Clark and his friend Andy.*

In the (a) version, the key word *Pete* serves as a subject NP, whereas in the (b) version, *Pete* serves as a direct object NP. Since this key word appears in approximately the same location within the two versions, any significant difference in F_0 may be attributed to the inherent effect of grammatical relation. The results revealed a very small, nonsignificantly higher value for the peak F_0 in *Pete* in Sentence (3b), where *Pete* appeared as an object. The mean value of the peak was 1.3 Hz higher in (3b) ($p > .20$). This finding suggests that NP subjects do not exhibit inherently higher F_0 peaks than objects. Thus, the results of the previous experiment for *P2* seem to reflect the fact that the *P2* word begins a new clause rather than that this word happens to be a subject. The absence of inherent F_0 differences between subjects and objects in this case further suggests that *P1* values typically lie well above the Topline Rule in our studies in Chapter 2 not by virtue of their subject status but rather as a result of their location as the first stressed syllable in the clause. The results of Experiment 2.3.4 have also shown that the *P1* word lies above the Topline Rule, regardless of whether this word is an Adjective or the head Noun of the subject NP, indicating that the grammatical category of the word is not a primary controlling influence.

To sum up, the results of this experiment indicate that fall–rise patterns are much more pronounced for two words that span the boundary between two clauses than when the same key words appear within the same phrase. These results confirm the general prediction that the magnitude of fall–rise patterns is greater at stronger syntactic boundaries. The greater fall in F_0 at the clause boundary is attributable to a much lower valley and an elevated *P2*, with *P1* remaining approximately constant at strong vs. weak boundaries.

Experiment 3.1.2[1]

In this experiment we sought to test a more fine-grained prediction of the boundary strength metric by determining whether the fall–rise pattern of F_0 occurring at the boundary between two main clauses is larger in magnitude than the pattern occurring between a main and embedded clause. According to the theory of boundary strengths, the magnitude of fall–rise pattern should be greater between the two main clauses.

Five males and four females participated in this experiment. Four pairs of sentences were constructed, as shown below with key words underlined. In each pair, the (a) version contained two conjoined main clauses, while the (b) version contained one main clause and one embedded clause. The embedded clause involved a complement structure in Sentences (4b) and (5b) and a restrictive relative structure in Sentences (6b) and (7b).

(4) a. *Anthony was surprised and Raymond became upset.*
    b. *Anthony was surprised Andrêa became upset.*
(5) a. *Almost everyone knows but Linda hasn't found out yet.*
    b. *Almost everyone knows Belinda hasn't found out yet.*
(6) a. *Marie was listening to the song and Del was playing.*
    b. *Marie was listening to the song Adelle was playing.*
(7) a. *Marie was listening to the song and Raymond was playing.*
    b. *Marie was listening to the song Andrêa was playing.*

The sentences of each pair were matched for total number of syllables and contained the same approximate stress contour (see later). In (4b) and (7b), the speakers were instructed to place stress on the second syllable of *Andrêa* in order to render a close match in stress with *and Raymond*. The structural difference between the (a) and (b) sentences can be captured in terms of the location of clause brackets. As shown in (8) below, the major utterance-internal syntactic boundary in the (a) version contains both a right S bracket and a left S bracket, representing the end and beginning of the two conjoined

(8) a. $_S[$*main clause*$]_S$ $_S[$*main clause*$]_S$
    b. $_S[$*main clause* $_S[$*embedded clause*$]_S]_S$

[1]Experiments 3.1.2 and 3.1.3 are revised from Cooper and Sorensen (1977). We thank the American Institute of Physics for permission to adapt this material.

clauses. The (b) version, on the other hand, contains only a single S bracket at the major boundary, representing the beginning of the embedded clause. The right S bracket denoting the end of the main clause in (b) occurs at the end of the entire string, consistent with the notion that the main clause properly contains the embedded clause in the structural description.

The embedded clause in the (b) sentences is referred to as a REDUCED CLAUSE because the complementizer or the relative pronoun *that* which normally introduces the clause is absent from the surface string. Such deletion was necessary to achieve a close degree of phonetic similarity between the (a) and (b) sentences of each pair in the region of the syntactic boundary. On linguistic grounds, it is conceivable that the deletion of *that* in the (b) sentences reduces the remaining embedded material to the status of a phrase (Hankamer, 1971) by operation of a convention known as tree pruning (Ross, 1969). This convention serves to delete nonbranching clause nodes from the surface structures. If so, the main internal syntactic boundary in the (b) sentences would contain no S brackets whatsoever, only phrase-level brackets. Regardless of whether the tree pruning convention is applied, however, the clause boundaries of the (a) sentences would be considered stronger than the key boundaries of (b) according to the proposed metric of boundary strengths (Cooper and Paccia-Cooper, 1980). Thus, larger fall–rise patterns are predicted in the (a) vs. (b) versions.

The F_0 measurements for this and the following experiment were obtained not from FPRD but from an interactive computer hardware and software package which is implemented on the M.I.T. Natural Language Processing Group's PDP-9 computer. With this algorithm, F_0 measurements are obtained from six parallel F_0 estimators, utilized as input to a final F_0 computation (Gold and Rabiner, 1969; O'Shaughnessy, 1974). A digitized waveform of the speech (sampling rate = 10kHz) is available to the user on a refresh-type display. The portion of speech for which an F_0 contour is desired can be located on this display, and an appropriate command replaces this waveform with its estimated F_0 contour. This contour is represented as a series of F_0 estimates having a spacing of 10 msec, and a cursor can be aligned with any point on the display to provide a read out of the F_0 value at that point. A sample F_0 display is shown in Figure 3.3.

The resolution of this F_0 algorithm is 80 μsec. Thus, varying degrees of quantization occur for F_0 values in different frequency regions (see Chapter 1). A signal containing F_0 values around 200 Hz for a typical female speaker is quantized at about 3 Hz. For male speakers, the quantization is typically between 1 and 2 Hz.

Following the procedure of Experiment 3.1.1, three F_0 measurements were recorded as data for each utterance in the region of the major syntactic boundary. The data included the peak value of F_0 in the stressed syllable just prior to the major boundary (*P1*), the lowest value in the same syllable (*V*), representing a valley, and the peak value in the first stressed syllable following the boundary (*P2*). Thus, for sentence (4b), for example, *P1* and *V* values were obtained for the syllable /praˈzd/ in *surprised,* and *P2* was obtained for the

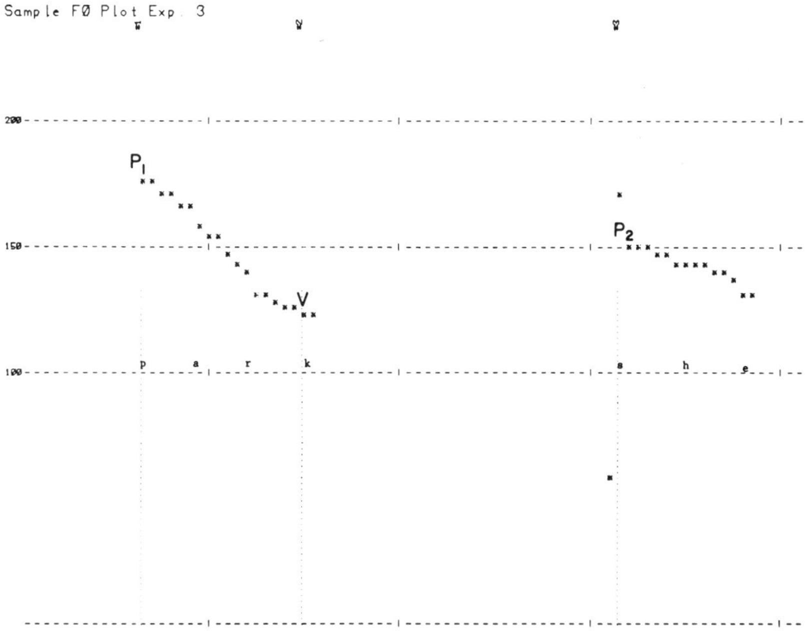

Figure 3.3. Sample plot of F_0 vs. time from the Gold and Rabiner (1969) pitch detector. This plot shows the F_0 fall–rise pattern for the words *park* and *she* in the utterance *The owner of the park she scolded the children,* involving Left Dislocation (see Experiment 3.1.3). The F_0 values for *P1*, *V*, and *P2* for this utterance, spoken by a male, were 176, 123, and 150 Hz, respectively.

syllable /dre/ in *Andréa.* To minimize the chances of including spurious F_0 estimates in the data, peak and valley measurements were adopted which represented two or more contiguous occurrences of the same F_0 value (see Figure 3.3). In cases where the F_0 contour in the region of interest was monotonically decreasing or increasing, the second and second-to-last points were used as data.

The mean values of each of the measured points *P1*, *V*, and *P2* are shown in Table 3.2 for each of the eight test sentences. Figure 3.4 shows the mean of *P1*, *V*, and *P2* for the (a) and (b) versions. In each sentence, the presence of a fall–rise pattern is apparent. The fall in F_0 from *P1* to *V* averaged 23.0 Hz in the conjoined clause sentences of (a) and 20.5 Hz in the main-embedded clause sentences of (b). The fall in F_0 from *Pl* to *V* was statistically significant in each of the eight sentences ($p < 0.01$, t values ranging from 4.52 to 8.69, $df = 8$). The rise in F_0 from *V* to *P2* averaged 20.9 Hz in the conjoined main clause sentences and 9.83 Hz in the main-embedded clause sentences. This F_0 rise was statistically significant in seven of the eight sentences ($p < 0.05$, t values ranging from 2.39 to 4.69, $df = 8$). The single exception was Sentence (7b), a main-embedded sentence, which showed a nearly significant effect in the same direction ($0.10 > p > 0.05$, $t = 2.00$, $df = 8$).

Table 3.2. Mean F_0 values for experiment 3.1.2

Sentence		*P1*	*V*	*P2*
(4)	a.	170	145	164
	b.	173	151	159
(5)	a.	174	144	180
	b.	174	152	166
(6)	a.	165	145	156
	b.	166	147	155
(7)	a.	165	147	165
	b.	168	149	158
Grand-mean	a.	169	145	166
	b.	170	150	160

Comparison of the magnitude of F_0 fall ($P1-V$) in the (a) and (b) versions of each sentence pair revealed a larger magnitude of fall in the (a) version of three of the four pairs. In Sentence Pair (5), this difference reached statistical significance ($p < 0.05$, $t = 2.35$, $df = 8$). With the F_0 rise ($P2-V$), a similar comparison of the magnitude in the sentence pairs revealed a larger magnitude of rise in the (a) version of each of the four sentence pairs. This difference also reached statistical significance for Sentence Pair (5) ($p < 0.01$, $t = 3.57$, $df = 8$). It thus appears that the fall–rise pattern of F_0 is slightly more pronounced at the boundary between two main clauses than at the boundary between a main and embedded clause. This overall difference in fall–rise pattern is evident in Figure 3.4. The greater magnitude of the fall–rise pattern at

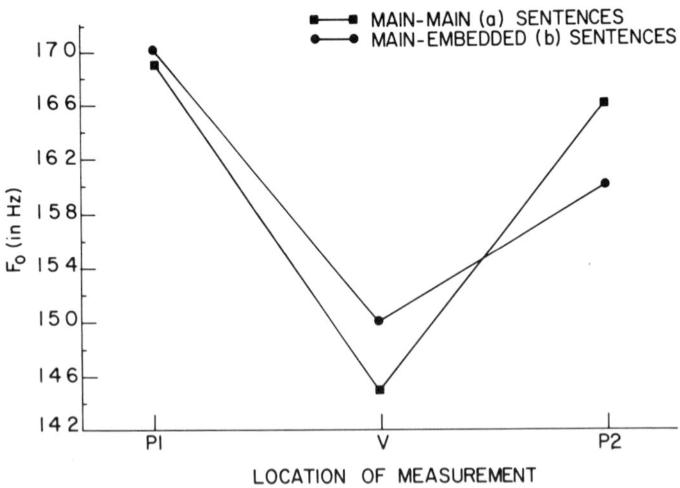

Figure 3.4. Plot of mean F_0 vs. time for each of the measured values in Sentences (4)–(7).

the boundary between the two main clauses is attributable both to somewhat lower values of *V* and higher values of *P2* in the (a) sentences. The difference between the fall–rise patterns for the (a) vs. (b) sentences was carried primarily by the F_0 rise (see Figure 3.4).

A comparison of the values for the individual points *P1*, *V*, and *P2* in each sentence pair (refer to Table 3.2) revealed that the values for *P1* in the (a) versions were equal or lower than the values for *P1* in the (b) versions in each of the four sentence pairs. The values for *V* in the (a) versions were lower than the values for the *V* in the (b) versions of each pair, but the values of *P2* were higher in the (a) versions. The only statistically significant effect observed for these comparisons was for *P2* in Sentence Pair (5) ($p < 0.01$, $t = 4.85$, $df = 8$).

The peak values *P1* and *P2* may be considered as estimates for the amount of stress placed on the key segments. Although some stress differences may not be manifested in these peak values, the absence of major differences in the (a) and (b) versions at the two peaks (see Table 3.2) suggests that the stress assigned to the key segments near the syntactic boundaries was approximately equal for the (a) and (b) versions (see also Experiment 3.1.3). If so, the difference in the magnitude of the fall–rise pattern for the (a) vs. (b) sentences can be attributed to the stronger syntactic boundary in (a) rather than to an intermediary influence of stress.

The values of *P1* and *P2* were also compared with each other in each sentence, to determine whether the fall–rise pattern represents a local inflection or a resetting of the F_0 declination function. As shown in Table 3.2, the average value for *P1* was larger than *P2* in six of the eight sentences and for each of the four (b) sentences. In three of the four (b) versions and in one of the (a) versions, the greater magnitude of *P1* over *P2* reached statistical significance ($p < 0.01$, t values ranging from 3.71 to 8.38, $df = 8$). The generally higher values for *P1* suggest that the fall–rise pattern observed at these syntactic boundaries represents a local F_0 inflection and that a resetting of the speaker's declination function did not occur (see Chapter 2). We suspect that the absence of resetting at the boundary between two conjoined clauses here is attributable to the relatively short length of each clause. When longer clauses were spoken, as in Study 2.4, resetting was observed. Further work is needed to determine whether resetting in such cases is triggered by the speaker's reaching some minimum F_0 value at the end of the first clause, by the speaker's anticipation of producing abnormally low F_0 values in the second clause if resetting is not accomplished, and/or by other factors, including secondary constraints on breathing (cf. Study 2.4).

Experiment 3.1.3

In this experiment, the study of fall–rise patterns of F_0 was extended to boundaries in sentences that departed from the normal Subject–Verb–Object order of English. These departures from SVO order appear to result from the application of syntactic transformations of preposing which move a constituent to the front of a sentence during its derivation (Emonds, 1976). These transformations typically serve to add focus to the moved constituent (Langacker, 1974).

In this experiment, we examined fall–rise patterns at the boundary between the fronted constituent and the remainder of the sentence. According to the account provided by Emonds (1976), this boundary is considered to exhibit the status of a major clause boundary. Accordingly, it is predicted that a greater fall–rise pattern should accompany such a boundary in comparison to the boundary between major NP and VP phrases within a major clause. It is expected that the difference between preposed and SVO sentences will be particularly manifest for the rising portion of the fall–rise pattern, since the difference in the strength of the phrase immediately following the key boundary is greater for the preposed vs. SVO sentences than is the difference in strength of the preceding phrase, as shown in Figure 3.5.

O'Shaughnessy (1976, p. 297) studied F_0 patterns for such sentence structures with four speakers and found that the end of the fronted constituent was

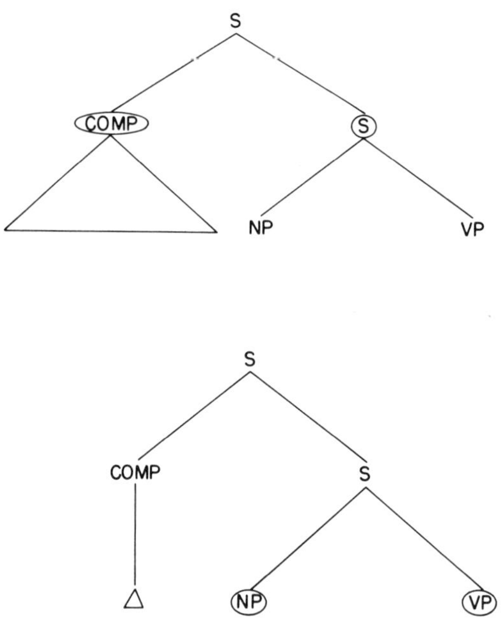

Figure 3.5. Schematic tree diagrams of SVO sentences and sentences involving preposing. Following Chomsky (1970), Emonds (1976), and Jackendoff (1977), it is assumed that the phrase structure rule S → (COMP) + S applies to all sentences, and that the complement node is occupied by the preposed phrase in the upper tree but is null in the lower tree. The circled nodes refer to the most inclusive nodes which border the key boundary in each case. While the phrase structure rule S → (COMP) + S is assumed to apply to all sentences described in this and other experiments, other figures do not include this abstract rule because it bears no consequences for the other structural distinctions at issue.

indeed marked by a larger fall in F_0 than a corresponding location in an SVO sentence. In addition, continuation rises were typically observed at the end of the fronted constituents. In this experiment, we extended the F_0 analysis of such sentence structures to cases in which the phonetic environments of the major syntactic boundaries were more closely matched with a larger group of speakers.

Two groups of four sentences were used in this experiment. In each group, one sentence contained normal SVO order, while the other three contained preposed constituents derived via the transformations of Left Dislocation, Topicalization, and Prepositional Phrase Preposing. Left Dislocation moves an NP to the front of a sentence and leaves a pronoun at the movement site, as in (9b); Topicalization moves a direct object NP to the front of a sentence, as in (9c); and PP Preposing moves a PP to the front, as in (9d). The sentences of Group (10) represent versions of the sentences in (9) in which the constituent preceding the major boundary was longer and slightly more complex. The sentences of each group were matched for total number of syllables and contained the same overall stress contour. However, given these diverse sentence structures, it was not possible to concurrently match the phonetic structure in the region of the boundary and the stress assigned to the word immediately following the boundary, namely, *shows, she('s),* or *Cher('s).* One can assume that the pronoun *she('s)* and the verb *shows* typically receive less stress than the proper noun *Cher('s).* The data suggest, however, that these stress differences do not lead to a principled account of the major results (see p. 116). All eight sentences appear below along with descriptive labels.

(9) a. *The owner of the park shows gold to the children.* SVO
b. *The owner of the park she scolded the children.* Left Dislocation
c. *The statue in the park Cher showed to the children.* Topicalization
d. *At Brockton's city park Cher scolded the children.* PP Preposing

(10) a. *The ragged old woman planting flowers in the park shows photographs to the children.* SVO
b. *The ragged old woman planting flowers in the park she's photographing the children.* Left Dislocation
c. *The ragged old woman planting flowers in the park Cher photographed with the children.* Topicalization
d. *Inside the old building full of flowers in the park Cher's photographing the children.* PP Preposing

According to the findings of O'Shaughnessy (1976), it was expected that the word *park,* occurring at the end of the preposed constituent in (9b–d) and (10b–d), would show a larger fall in F_0 than when *park* occurs at the end of the subject NP in (9a) and (10a).

Six males and four females participated in this experiment. Nine of the subjects also served in Experiment 3.1.2.

The testing procedure was similar to that of Experiment 3.1.2. F_0 values were obtained for *P1* and *V* in the constituent-final word *park* and for *P2* in

Table 3.3. Mean F_0 values for experiment 3.1.3

Sentence		*P1*	*V*	*P2*
9)	a.	178	149	164
	b.	173	143	160
	c.	170	139	169
	d.	165	137	165
(10)	a.	157	132	151
	b.	165	130	158
	c.	162	132	156
	d.	158	131	160

the next word, beginning with /š/ [*shows, she('s), Cher('s)*]. *P2* typically occurred during the first 50 msec of the postboundary word. The mean values of *P1*, *V*, and *P2* for each sentence are shown in Table 3.3. The data for the sentence (9) versions are graphed in Figure 3.6; for Sentence (10), in Figure 3.7.

The fall in F_0 for *park* (*P1–V*) averaged from 25 Hz in (10a) to 35 Hz in (10b), and this fall was statistically significant for each of the eight sentences ($p < 0.01$, t values ranging from 4.11 to 8.09, $df = 9$). Although the average fall was slightly larger for the preposed constituents in (9b–d) and (10b–d) than in the SVO sentences, these differences did not reach statistical significance ($p > 0.05$, t values ranging from -0.14 to 2.02, $df = 9$). The results with phonetically matched sentences thus do not support the notion that a reliably greater fall in F_0 accompanies the end of a preposed constituent (cf. O'Shaughnessy, 1976).

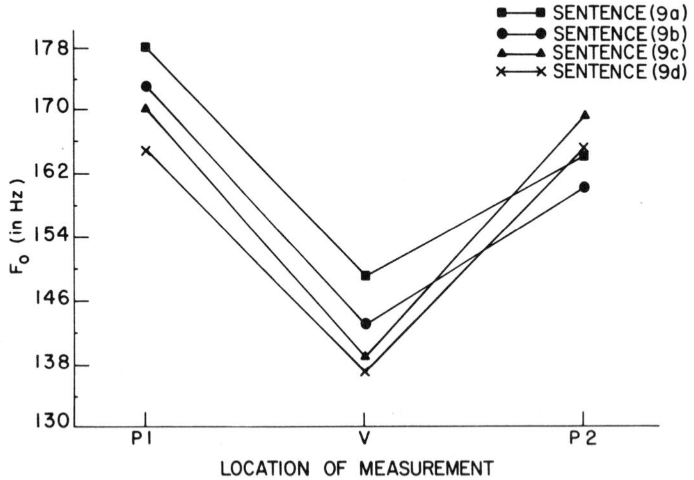

Figure 3.6. Plot of mean F_0 vs. time for each of the measured values in Sentence (9).

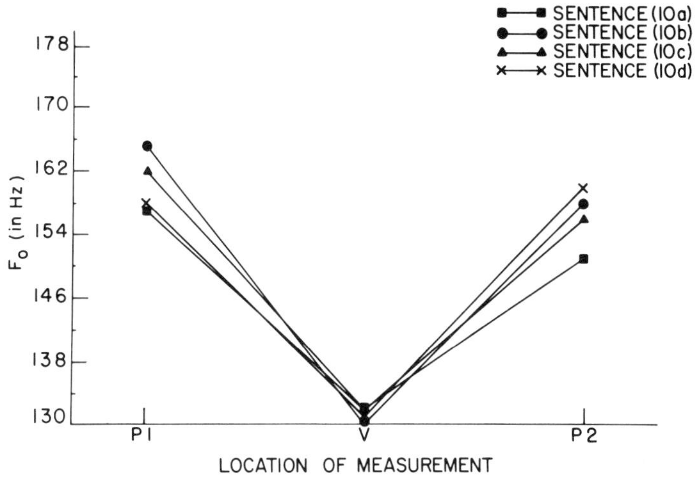

Figure 3.7. Plot of mean F_0 vs. time for each of the measured values in Sentence (10).

The rise in F_0 following the phrase boundary (*P2–V*) averaged from 15 Hz in (9a) to 30 Hz in (9c). This rise was statistically significant for each of the eight sentences ($p < 0.02$, t values ranging from 2.94 to 6.28, $df = 9$). The average rise in F_0 was greater in all six cases of preposing (9b–d) and (10b–d) than in the corresponding SVO sentences (9a) and (10a). In four cases of preposing, this difference was statistically significant: (9a) vs. (9c), $p < 0.02$, $t = 2.96$, $df = 9$; (9a) vs. (9d), $p < 0.02$, $t = 2.83$, $df = 9$; (10a) vs. (10b), $p < 0.05$, $t = 2.43$, $df = 9$; (10a) vs. (10d), $p < 0.05$, $t = 2.41$, $df = 9$. Based on these results, it appears that preposing exerts a greater and more consistent effect on the rise in F_0 following the fronted constituent than on the fall in F_0 at the end of this same phrase. This asymmetry may be accounted for in terms of the differences in surface structure at the boundary of the preposed vs. SVO sentences. As noted earlier, in the cases of preposing, this boundary is immediately followed by a Sentence node, marking the beginning of the main clause (Emonds, 1976). Thus, the difference in strength between the preposing vs. SVO sentences is most apparent for the phrase nodes just following the boundary rather than just preceding it. This difference in left–right strength accounts for the fact that a greater difference between preposing and SVO was observed for the rise rather than for the fall component of the fall–rise pattern in this experiment.

The rise vs. fall asymmetry was also observed in the earlier study of preposing by O'Shaughnessy (1976), involving four speakers. In that study, the greater rise with preposing appeared most prominently as a continuation rise on the syllable immediately preceding the boundary. Informal observation suggests that the greater magnitude of rise observed in our experiment was also attributable in large part to such a continuation rise. Although such a pattern appears just prior to the key boundary, its magnitude seems to be directly influenced by the strength of the phrase immediately following the boundary, in accordance with the interpretation provided above.

Comparison of the values for each of the measured points *P1*, *V*, and *P2* revealed no highly systematic differences across versions (a)–(d) in Sentence Groups (9) and (10). In Group (9), however, *P1* in (9a) was significantly higher than *P1* in (9d), and *V* in (9a) was significantly higher than *V* in (9b–d) ($p < 0.05$, t values ranging from 2.78 to 4.79, $df = 9$). In Group (10), *P2* of (10a) was significantly lower than *P2* of (10d) ($p < 0.05$, $t = 2.95$, $df = 9$).

It is clear from these results that possible differences in the stress assigned to *shows, she('s),* and *Cher('s)* cannot account for the differences noted previously for the magnitude of F_0 rise in SVO vs. preposed structures. In particular, examination of the values of *P2* for Sentences (9a–d) in Table 3.3 shows no systematic relation between *P2* and the magnitude of F_0 rise (*P2*–*V*), and the difference in *P2* for (10a) vs. (10d) noted above cannot lead to a unified account of the structural effect on F_0 rise.

Although values of *P2* provide an estimate of stress level for *shows, she('s),* and *Cher('s)* it must also be recognized that stress differences may have occurred which are not reflected in these values. The most plausible stress differences, as noted previously, would be the assignment of greater stress on *Cher('s)* than on *she('s)* and *shows*. However, any account based on this difference also fails to provide a rationale for the structural effect on F_0 rise, since the rise was observed to a greater extent in the sentences containing *she('s)* as well as *Cher('s)*. In conclusion, it appears that the amount of F_0 rise can be attributed to a structural difference, as noted above, rather than to differences in stress level.

Comparison of the value of each measured point for the short and long constituents of (9) vs. (10) for each sentence (a)–(d) reveals that *P1*, *V*, and *P2* were always higher in (9) than in (10) (see Table 3.3). The consistently lower F_0 values in the key region of (10) are predicted by the notion that F_0 inflections take on lower values for each successive syllable in a constituent, due to the influence of F_0 declination (see Chapter 2).

Comparison of the values of *P1* and *P2* revealed that the average values of *P1* were larger in seven of the eight sentences, as shown in Table 3.3. In two sentences, this difference reached statistical significance [(9a), $p < 0.01$, $t = 4.39$, $df = 9$; (9b), $p < 0.01$, $t = 3.36$, $df = 9$]. The results suggest that the fall–rise pattern observed at phrase boundaries in this experiment represents a local F_0 inflection without the presence of a resetting of the F_0 declination function, similar to the effects observed at clause boundaries in Experiment 3.1.2.

In earlier work, Maeda (1976) obtained instances of both local inflection and resetting at the boundary of a preposed prepositional phrase (Maeda, 1976, Figs. 2.1a–2.1c). The present data show, however, that the local inflection occurs with greater consistency.

The results of this experiment were compared with data on the duration of the word segment /par/ in *park* for the same utterances (Cooper and Paccia-

Cooper, 1980). In the study of speech timing, the duration of this word segment was significantly longer when it appeared at the end of a preposed constituent. However, we found no significant correlations between segment duration and magnitude of F_0 fall from *P1* to *V* in this word ($p > 0.05$ in each case; two-tailed Spearman's rank correlation). The absence of such correlations casts some doubt on the notion that segmental lengthening is produced by the speaker in order to achieve a greater amount of F_0 fall (see Chapter 5).

To summarize, the results of this experiment generally support the idea that a major clause boundary appears just after the end of a preposed constituent. As in the previous two experiments, greater magnitudes of fall–rise patterns were observed at clause boundaries following each of three different types of preposed constituent than at boundaries considered to be weaker according to the metric of boundary strengths (Cooper and Paccia-Cooper, 1980). Here, the major effects of the clause boundary appeared on the F_0 rise rather than on the F_0 fall, suggesting that the effect of preposing is primarily attributable to the beginning of an upcoming clause immediately following the preposed constituent.

Conclusion

The results of the past three experiments indicate that a local fall–rise pattern of F_0 accompanies the boundaries of clauses in a variety of sentence types. In addition, the magnitude of these fall–rise patterns was greater at strong boundaries than at boundaries predicted to be weaker on the basis of the metric of boundary strength. The findings thus provide some support for the idea that the magnitude of a fall–rise pattern at a given syntactic boundary is directly related to the independently specified strength of that boundary.

The results of this study have also been helpful in distinguishing the individual components of boundary strength for the phrases immediately preceding and following the boundary. With preposing, the magnitude of the rising portion of the fall–rise pattern was increased consistently more than the magnitude of the falling portion, in agreement with the fact that the major increment in boundary strength at the boundary between the preposed constituent and the rest of the sentence is attributable to the presence of an upcoming main clause just following the boundary.

Study 3.2. Phrase Boundaries

We now turn to examine fall–rise patterns at the boundaries of major phrases within a single main clause. By so doing, it will be possible to determine in some detail the way in which the speaker's coding of phrase structures exerts a systematic influence on F_0 patterns.

In formal linguistics, the phrase structure rules of a language are formulated as a set of rewrite rules of the form "*X* → *Y*" (rewrite constituent(s) "*X*" as constituent(s) "*Y*"). These rules specify both the hierarchical and linear relationships among the constituents in a sentence (e.g. Chomsky, 1957, 1965). In addition, the rules provide an algorithm for deriving a tree diagram in which the structural representation of the sentence is more readily comprehended (see below).

By examining the nature of rewrite rules, it is possible to formulate issues about the speaker's coding of phrase structure. For example, let us consider the structural representation for a simple sentence: *Jeanne and Linda like the countryside*. The rewrite rules for generating this sentence appear below in (11):

(11) (1) S → NP + VP
(2) NP → N + CONJ + N
(3) N → Jeanne
(4) CONJ → and
(5) N → Linda
(6) VP → V + NP
(7) V → like
(8) NP → DET + N
(9) DET → the
(10) N → countryside

As shown in (11), the system of rewrite rules is quite intricate for even a simple sentence. These rewrite rules are applied in a specific order, according to the top-down, left-to-right method of application shown in (12):

(12) (1) S
(2) NP + VP
(3) N + CONJ + N + VP
(4) N + CONJ + N + V + NP
(5) N + CONJ + N + V + DET + N
(6) Jeanne + CONJ + N + V + DET + N
(7) Jeanne and + N + V + DET + N
(8) Jeanne and Linda + V + DET + N
(9) Jeanne and Linda like + DET + N
(10) Jeanne and Linda like the + N
(11) Jeanne and Linda like the countryside

This structural representation can be recast as a tree diagram, shown in Figure 3.8.

In grade school, one is typically presented with some method of diagramming the structural representation of sentences, but seldom is one aware of the major theoretical assumptions and implications of the system that has been adopted. Why, for example, do we assume the existence of a constituent like VP which dominates the constituents V + NP in a sentence like (12)? Current studies in linguistic theory are devoted to questions such as this, and the outcome of such studies sometimes leads to major revisions in the form of rewrite rules.

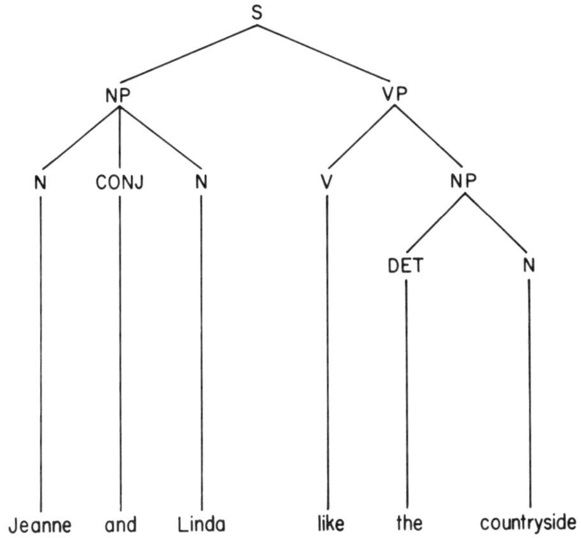

Figure 3.8. Tree diagram corresponding to the set of rewrite rules in (11).

From a psychological standpoint, the system of rewrite rules proposed on linguistic grounds leads us to naturally ask whether the relationships among constituents in speech coding are analogous to those specified in the formal grammar. This question can be partitioned as follows:

(a) Do speakers compute constituents like Noun Phrase and Verb Phrase as units?
(b) Do speakers code constituents hierarchically according to the system of rewrite rules proposed in linguistic theory?

In a study of speech timing, these questions were investigated in a variety of structurally ambiguous sentences (Cooper, Paccia, and Lapointe, 1978; Cooper and Paccia-Cooper, 1980). The results generally provided affirmative answers to each of the two questions raised above. In this study, we decided to analyze F_0 patterns for a number of the same utterances in order to obtain information about how the speaker's syntactic code influences F_0.

In each of the following experiments, two different readings of a structural ambiguity can be captured in terms of a difference in the strength of the syntactic boundary that coincides with the end of the first key word. In the study of speech timing, longer durations were obtained for the reading in which a stronger boundary acccompanied this key word, in agreement with the theory of boundary strengths. For these structures, the difference in strength is conveyed solely by the phrase immediately preceding the key boundary, and thus, unlike the previous study of preposing, it is expected that differences in fall–rise patterns of F_0 for these structures will be produced for the falling portion of this pattern but not for the rising portion. Previous observations of F_0 have already provided some indication of this pattern of results (e.g., Hartvigson,

1965; Lieberman, 1967; Crystal, 1969; Isačenko and Schädlich, 1970; Bolinger, 1972; Liberman, 1975; O'Shaughnessy, 1976, 1979). Since the predictions of the theory are best understood with reference to particular sentence structures, let us turn directly to the experiments themselves.

Experiment 3.2.1.

This experiment involved two ambiguous sentences:

(13) *My Uncle Abraham presented his talk naturally.*
(14) *Lieutenant Baker instructed the troop naturally.*

In each sentence, the adverb *naturally* modifies either the entire sentence or the Verb Phrase only, as shown in Figure 3.9. Figure 3.9 indicates that the key word *talk* marks the end of two phrases in the upper reading, including both the direct object NP and the superordinate VP, whereas this same key word marks the end of the direct object NP alone in the lower reading. The difference in strength is carried solely by the phrase immediately preceding the key boundary, since, as shown in Figure 3.9, the branching depth of the material following the key boundary is constant. Thus, it is expected that the difference in strength will produce a greater fall in F_0 at this boundary without producing a greater F_0 rise. The two meanings conveyed by this structural distinction can be paraphrased by (15a) and (15b) below.

(15) a. *Of course my Uncle Abraham presented his talk.*
b. *My Uncle Abraham presented his talk in a natural manner.*

In the earlier study of timing, speakers were presented the sentences along with their intended meanings, as given below:

(13) a. *My Uncle Abraham presented his <u>talk</u> <u>naturally</u>.*
(Of course Abraham presented his talk)
b. *My Uncle Abraham presented his <u>talk</u> <u>naturally</u>.*
(Abraham presented his talk in a natural way)
(14) a. *Lieutenant Baker instructed the <u>troop</u> <u>naturally</u>.*
(Of course Lieutenant Baker instructed the troop)
b. *Lieutenant Baker instructed the <u>troop</u> <u>naturally</u>.*
(The Lieutenant instructed in a natural way)

The speakers easily recognized the distinct meanings of each ambiguous sentence. They were asked to read these sentences and fillers in a pseudorandom order following the testing procedure used throughout this project.

Acoustical measurements were obtained for the duration of the key segment /taʷ/ of *talk* in Sentence (13) and the key segment /tru/ of *troop* in Sentence (14), as well as for the pauses following these segments. The results for ten

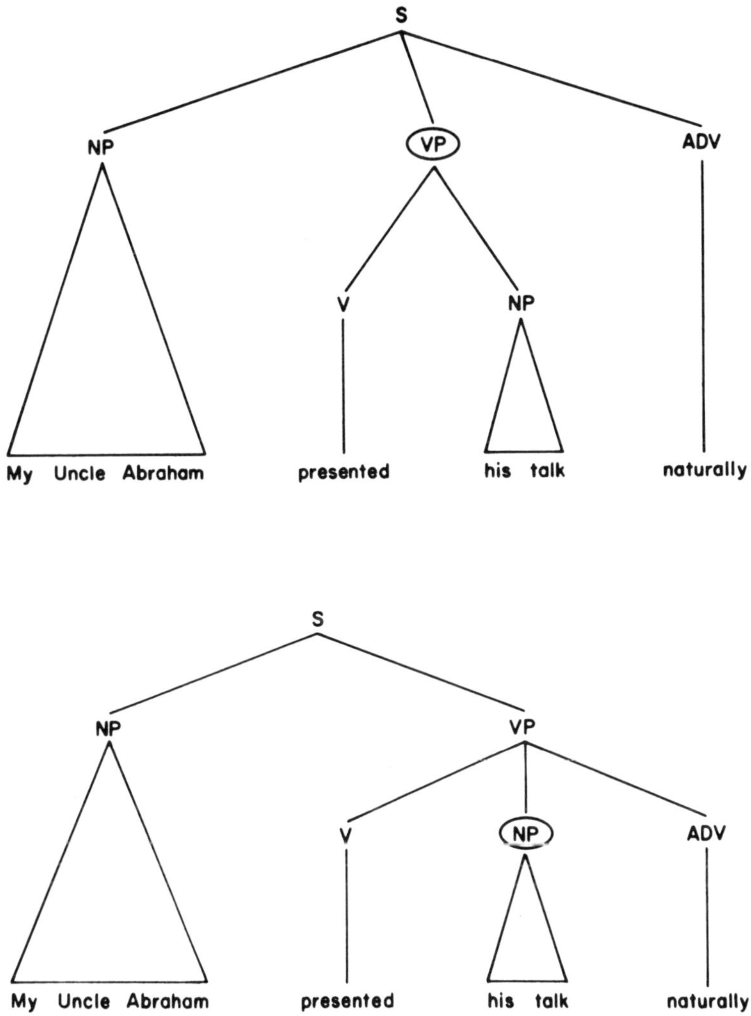

Figure 3.9. Structural representations of two readings of Sentence (13). A circle appears around the maximal constituent to the immediate left of the key boundary. The upper tree represents the reading in which *naturally* modifies the entire sentence; the lower tree represents the reading in which this adverb modifies only the Verb Phrase.

speakers showed that speakers lengthened the word segment and pause for the readings in which the key word coincided with the end of two phrases—the direct object NP and the superordinate VP—rather than when the key word coincided with the end of the direct object NP alone.

In this study, we examined fall–rise patterns of F_0 by measuring the peak (*P1*) and valley (*V*) of the key word segment [*talk* in (13) or *troop* in (14)] and the peak (*P2*) in the first stressed syllable of the following word *naturally*. These measurements were obtained for the utterances of seven speakers who

Table 3.4. Mean F_0 values for experiment 3.2.1

Sentence		*P1*	*V*	*P2*
(13)	a.	178	137	150
	b.	164	142	155
(14)	a.	176	134	168
	b.	169	152	153

were predetermined to produce the utterances with the least amount of emphatic stress. The mean values and standard deviations for *P1*, *V*, and *P2* for each sentence are presented in Table 3.4. Data for Pair (13) are shown in Figure 3.10, and for Pair (14) in Figure 3.11.

Within-Sentence Comparisons

Comparison of *P1* vs. *V* revealed a significant fall in F_0 for the first key word in each of the four sentence readings ($p < .05$, t values ranging from 2.62 to 4.66, $df = 6$). Comparison of *V* vs. *P2* showed nonsignificant F_0 rises for each of the four sentence readings ($p > .05$, t values ranging from 0.07 to 2.37, $df = 6$). Finally, comparison of *P1* vs. *P2* revealed significantly lower values of *P2* for three of the four readings, namely, (13a), (13b), and (14b) ($p < .05$, t values ranging from 2.78 to 4.53, $df = 6$), and a nonsignificant trend in the same direction for Sentence (14a) ($p > .20$, $t = 0.49$, $df = 6$). These results suggest that the fall–rise patterns observed in this experiment were produced during the domain of a single F_0 declination. This finding is in accord with the principle, discussed in Chapter 2, that speakers seldom begin a new domain of F_0 declination within the confines of a single main clause.

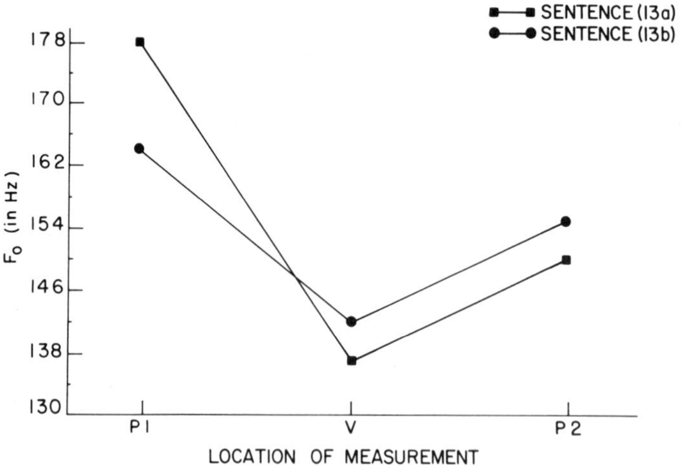

Figure 3.10. Plot of mean F_0 vs. time for the measured values of Sentence Pair (13).

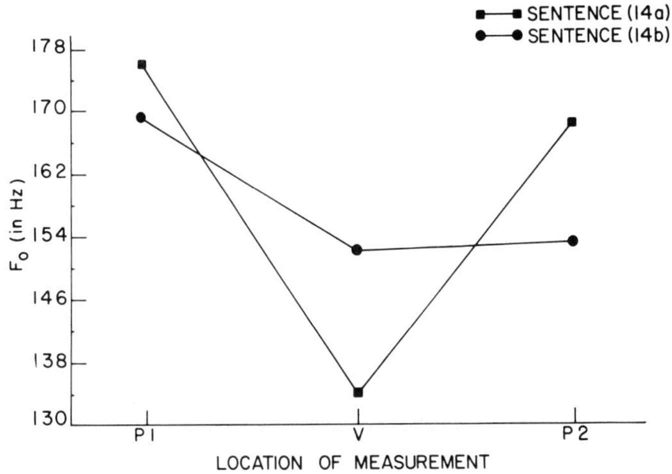

Figure 3.11. Plot of mean F_0 vs. time for the measured values of Sentence Pair (14).

Cross-Sentence Comparisons

The amount of fall (*P1–V*) in the (a) vs. (b) versions of both pairs was significantly greater for the (a) readings ($p < .02$, t values of 3.69 and 7.56, $df = 6$). This result suggests that the key word was accompanied by a more dramatic F_0 fall when this word marked the end of two phrases vs. one. As indicated in Figures 3.10 and 3.11, this greater F_0 fall is attributable to both higher values of *P1* and lower values of *V* in the (a) readings. There were no significant differences in the amount of rise (*P2–V*) in either pair, although the average magnitude of the rise in (14a) was considerable.

In conclusion, the results of this experiment indicate that a significant F_0 fall accompanies the end of an NP phrase as well as the end of a VP phrase. The magnitude of the F_0 fall, however, is significantly greater at the latter location, in which both the VP and NP phrases coincide. The results thus suggest that the addition of a single phrase node is sufficient to produce an increment in the amount of F_0 fall at a phrase boundary.

The findings of this experiment are in agreement with the predictions of the theory of boundary strengths. Since the only difference in boundary strength for the members of each sentence pair was conveyed by a phrase node just preceding the key boundary, a greater amount of fall in F_0 accompanied the reading with the stronger boundary, whereas the amount of rise in F_0 did not differ significantly in the two readings.

Experiment 3.2.2

In this experiment, the analysis of F_0 patterns was extended to a related type of structural ambiguity. In this case, the ambiguity hinges on a distinction in modification at a somewhat lower level in the structural representation of

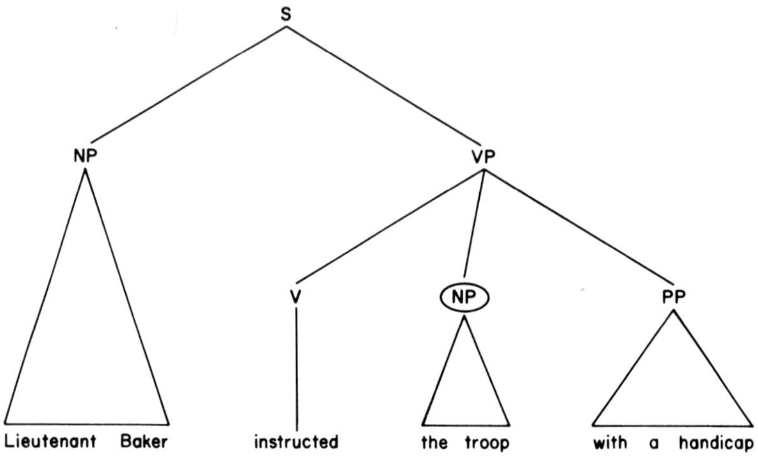

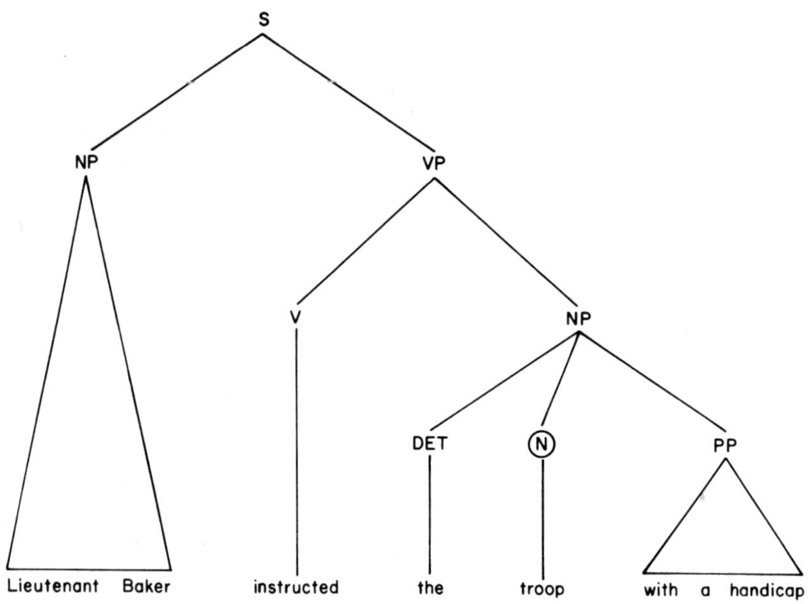

Figure 3.12. Structural representations of two readings of Sentence (16). The upper and lower trees correspond to the (a) and (b) readings, respectively, as described in the text.

the sentence than in Experiment 3.2.1, allowing one to determine whether the effects noted in the previous experiment are restricted to cases in which the modifiee is attached to the highest S in the tree.

The experiment involved two ambiguous sentences:

(16) *Lieutenant Baker instructed the troop with a handicap.*
(17) *Jeffrey hit the cop with a stick.*

In each sentence, the prepositional phrase beginning with *with* modifies either the Verb Phrase or the Noun Phrase dominated by the Verb Phrase, as shown for (16) in Figure 3.12. In (16) for example, the prepositional phrase can indicate either the manner in which Lieutenant Baker instructed the troop (Verb Phrase modification) or the condition of the troop itself (Noun Phrase modification). Figure 3.11 indicates that the key word *troop* is accompanied by a stronger boundary in the (a) vs. (b) reading. Again, the structural difference is conveyed exclusively by a difference in the strength of the phrase immediately preceding the key boundary, with the branching depth of the phrase immediately following this boundary held constant.

Ten subjects were presented with the sentences and disambiguating phrases, as below:

(16) a. *Lieutenant Baker instructed the troop with a handicap.*
(The Lieutenant was handicapped)
b. *Lieutenant Baker instructed the troop with a handicap.*
(The troop was handicapped)
(17) a. *Jeffrey hit the cop with a stick*
(Jeffrey had the stick)
b. *Jeffrey hit the cop with a stick.*
(The cop had the stick)

The speech samples for one of the original subjects was too low in amplitude to be suitable for F_0 analysis. Measurements of F_0 for *P1* and *V* for the remaining nine speakers were obtained for the key words *troop* (16) and *cop* (17). No measurements of *P2* were obtained, since the next word *(with)* was always unstressed. The mean values for the group of speakers appear in Table 3.5.

Table 3.5. Mean F_0 values for experiment 3.2.2

Sentence		*P1*	*V*
(16)	a.	170	142
	b.	172	162
(17)	a.	170	143
	b.	173	157

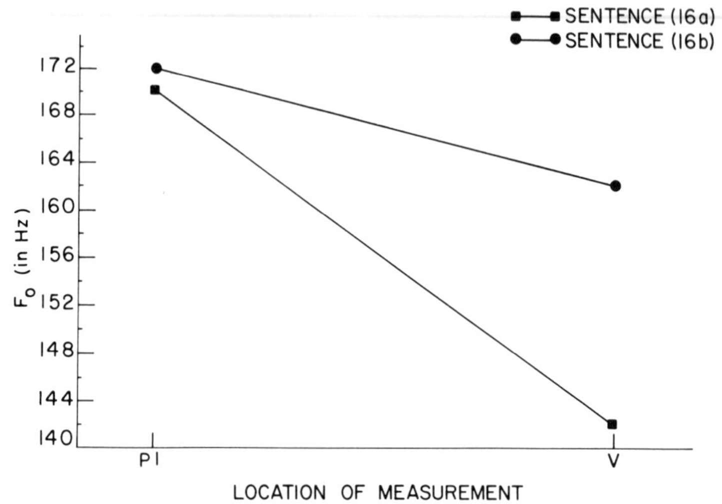

Figure 3.13. Plot of mean F_0 vs. time for the measured values of Sentence Pair (16).

These data are shown graphically in Figures 3.13 and 3.14 for Pairs (15) and (16), respectively.

Comparison of *P1* vs. *V* indicated the presence of a significant F_0 fall in Sentences (16a), (16b), and (17a) ($p < .005$, t values ranging from 4.05 to 4.69, $df = 8$) and a nearly significant fall in Sentence (17b) ($p < .10$, $t = 1.96$, $df = 8$). A comparison of the amount of F_0 fall in the two different readings of each ambiguity revealed a more pronounced fall in the readings in which the prepositional phrase modified the entire Verb Phrase. This effect was

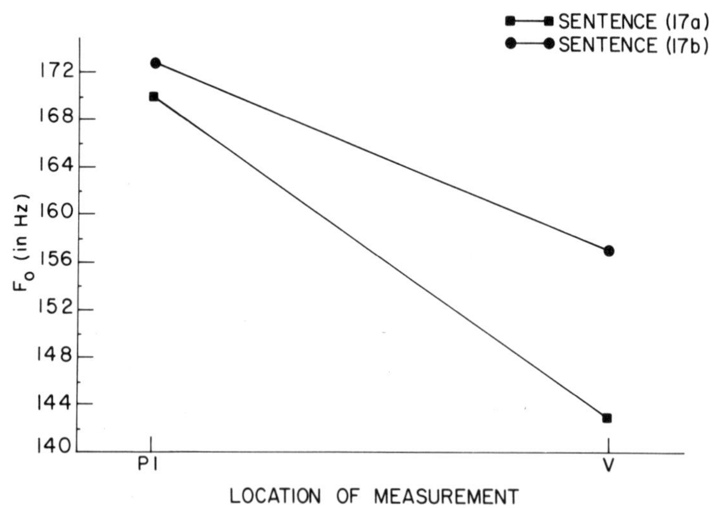

Figure 3.14. Plot of mean F_0 vs. time for the measured values of Sentence Pair (17).

significant in both Sentence Pair (16) and Sentence Pair (17) [Pair (16), $p < .02$, $t = 2.97$, $df = 8$; Pair (17), $p < .05$, $t = 2.31$, $df = 8$]. As indicated in Figures 3.13 and 3.14, this difference was primarily attributable to the lower values of V in the (a) versions. As in the previous experiment, a greater F_0 fall was observed for the reading in which the key word was accompanied by a stronger syntactic boundary. In this case, the effect was observed for a structural ambiguity in which the reading with the more inclusive key phrase boundary coincided with the end of an NP rather than a VP as in the previous experiment (see Figure 3.11), demonstrating that the effect is not restricted to a specific type of phrase node. The findings of both of these experiments are particularly noteworthy because significant effects were obtained for different readings of structural ambiguities distinguished by a single node of boundary strength. The data thus indicate that the magnitudes of F_0 falls are sensitive to relatively fine-grained structural distinctions.

Experiment 3.2.3

In order to further test the generality of the previous results, we measured fall–rise patterns of F_0 for another pair of structural ambiguities that had been used in the earlier study of speech timing. In this case, the ambiguity involved a distinction between two uses of the same Noun as either a vocative or as part of a compound Noun.

The two ambiguous sentences include:

(18) *Here is the famous Duke James.*
(19) *Here is the famous Pope Carl.*

In each sentence, the proper Noun *James* or *Carl* serves as either a Vocative, referring to the person to which the speech is directed, or as the second word of a compound proper Noun, *Duke James* or *Pope Carl.* The structural distinction between these two readings is represented in Figure 3.15. As in the previous experiments, the key words *Duke* and *Pope* were accompanied by a stronger syntactic boundary in the (a) readings, with the difference in boundary strength conveyed solely by the phrase immediately preceding the key boundary. The test sentences appeared with their disambiguated paraphrases as below.

(18) a. *Here is the famous Duke James.*
(James meets the Duke)
b. *Here is the famous Duke James.*
(The name of the Duke is James)
(19) a. *Here is the famous Pope Carl.*
(Carl meets the Pope)
b. *Here is the famous Pope Carl.*
(The name of the Pope is Carl)

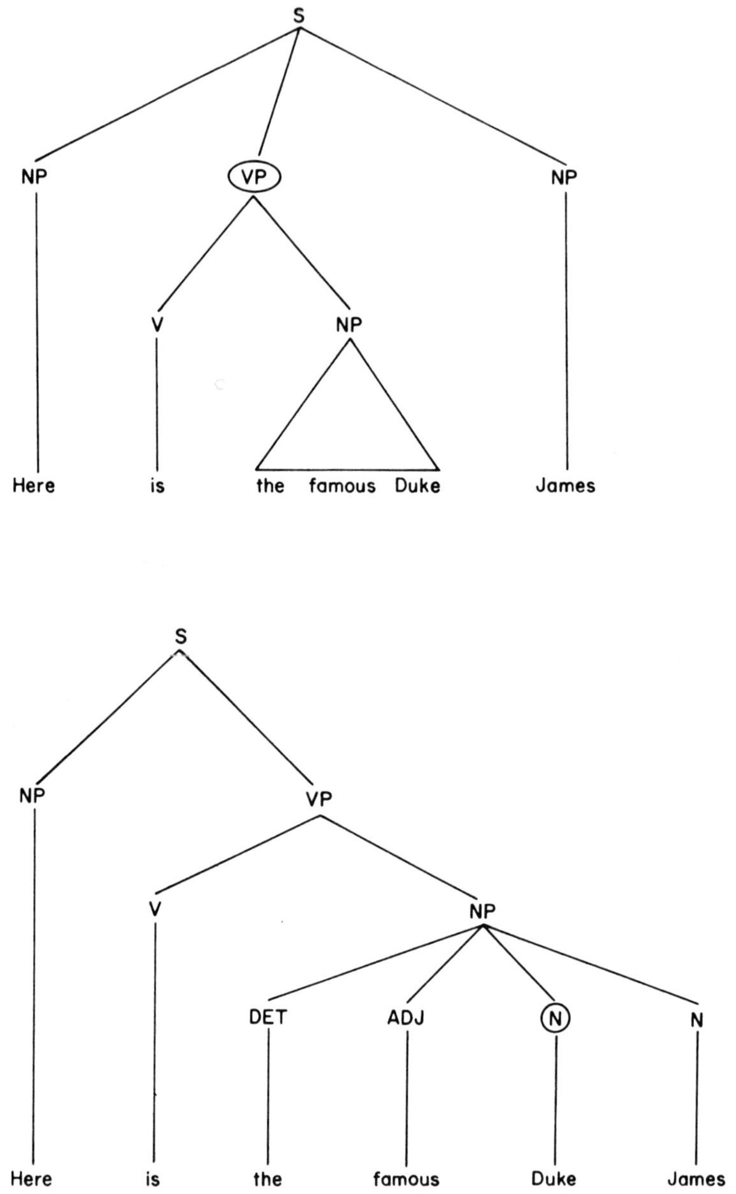

Figure 3.15. Structural representations of two readings of Sentence (18). The upper and lower trees correspond to the (a) and (b) readings, respectively, as described in the text.

Table 3.6. Mean F_0 values for experiment 3.2.3

Sentence		*P1*	*V*	*P2*
(18)	a.	194	174	162
	b.	184	178	175
(19)	a.	206	180	161
	b.	185	171	178

In the previous study of speech timing, acoustical measurements of duration were taken for the key segment /du/ of *Duke* and for /po/ of *Pope,* as well as for the following pauses in each sentence. Here, measurements of F_0 were taken for *P1* and *V* on the key words *Duke* or *Pope* and for *P2* on the following words *James* or *Carl.*

The utterances were measured for all 10 speakers who served in this experiment. The mean values for *P1*, *V*, and *P2* are presented in Table 3.6. The data for Pair (18) are plotted in Figure 3.16 and for Pair (19) in Figure 3.17.

Within-Sentence Comparisons

As expected, comparison of *P1* vs. *V* revealed a significant F_0 fall for the key word in each of the four sentence readings ($p < .02$, t values ranging from 3.14 to 4.85, $df = 9$). Surprisingly, however, comparison of *V* vs. *P2* also showed a significant FALL in F_0 from *V* to *P2* in each of the (a) readings ($p < .05$, t values of 2.46 and 3.56, $df = 9$), and a nonsignificant fall in (18b).

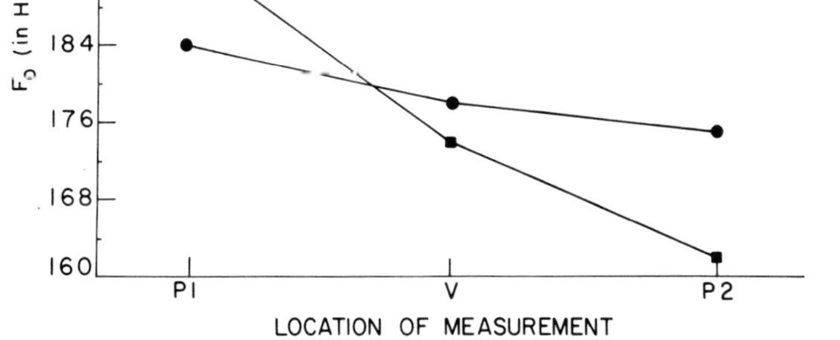

Figure 3.16. Plot of mean F_0 vs. time for the measured values of Sentence Pair (18).

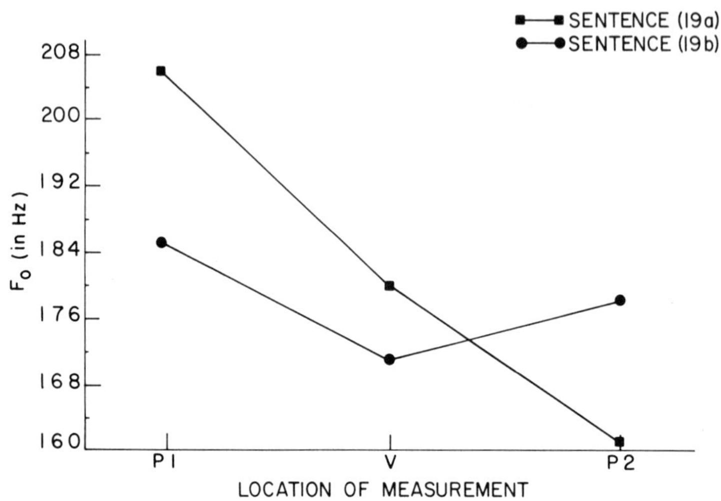

Figure 3.17. Plot of mean F_0 vs. time for the measured values of Sentence Pair (19).

Thus, the values of the first key word's valley were actually higher than the values of the peak of the following word in these sentences, as shown in Figures 3.16 and 3.17 (see below). Finally, comparison of *P1* vs. *P2* showed significantly lower values of *P2* in three of the four sentences ($p < .02$, t values ranging from 3.20 to 6.62, $df = 9$). In (19b), a nonsignificant trend was observed in the same direction ($.20 > p > .10$, $t = 1.47$, $df = 9$).

Cross-Sentence Comparisons

Comparison of the values for the (a) vs. (b) versions of each sentence revealed striking differences in the entire shape of the fall–rise patterns. To illustrate these differences, the fall–rise patterns of the (a) vs. (b) versions are schematized in Figure 3.18.

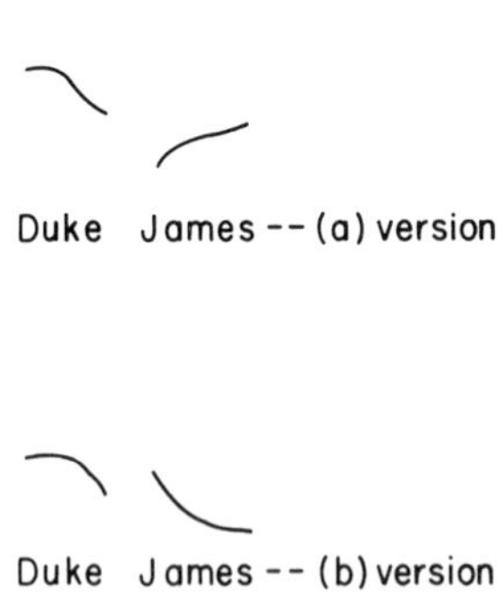

Figure 3.18. Schematic F_0 patterns showing the different types of fall–rise patterns observed in the (a) vs. (b) sentences of Experiment 3.2.3.

In the (a) versions, speakers typically exhibited a rising F_0 pattern on the key words *James* and *Carl,* such that *P2* occurred at the end of these words rather than at the beginning, as in the (b) version. These two key words typically began with very low values of F_0 in (a), representing a dramatic drop of about 30 Hz from the valley value at the end of the preceding key word. About half the F_0 "fall" was thus accomplished by a sudden drop in F_0 between the first and second word rather than exclusively by a fall within the first key word itself, although this key word did exhibit a fall typically amounting to about 20 Hz. The speaker thus employed a different means of achieving a dramatic fall–rise pattern in the (a) versions of these sentences than has typically been observed in our studies of such patterns. In particular, a major portion of the F_0 fall was accomplished by a sudden drop in F_0 across the word boundary, while the F_0 rise was primarily accomplished by a rising F_0 pattern within the second key word. The general pattern of results was obtained for similar ambiguous sentences by O'Shaughnessy (1979). In the other studies of fall–rise patterns in this chapter, however, the situation was just the opposite—the fall was primarily accomplished by a falling contour within the first key word, while the rise was primarily accomplished across the subsequent word boundary. Because the actual pattern of fall–rise pattern differed so greatly for the (a) and (b) versions of these sentences, it is not appropriate to consider statistical comparisons of the individual peak and valleys across the different sentences of each pair. The F_0 pattern exhibited in the (a) versions seems attributable to the presence of a vocative following the key word, since vocatives typically contain a rising F_0 (O'Shaughnessy, 1979). If so, then the unusual fall–rise patterns exhibited may be attributed chiefly to an independent influence of a vocative rather than to the strength of the syntactic boundary per se. Although this interpretation is completely ad hoc in the present case, it seems plausible in light of the fact that a vocative represents a completely different type of sentence mode from the declarative sentences examined in each of the previous experiments in this chapter.

Experiment 3.2.4

In this experiment, we sought to determine whether the previous results would generalize to paragraph contexts in which structurally ambiguous sentences were effectively disambiguated by prior linguistic context. In this way, it would be possible to assess whether the previous results represented a conscious effort on the speaker's part to disambiguate the two readings of a given ambiguity by F_0 pattern or whether the differences in F_0 reflect an influence that accompanies speech production under more normal circumstances as well. In the earlier study of speech timing, the same sort of effects on segmental duration and pausing were observed in disambiguating paragraph contexts, although the consistency and magnitude of these effects were not as strong as in isolated sentences.

In this experiment, we examined the structural ambiguity for which F_0 measurements had already been obtained in isolated sentences in the preceding experiment. For each reading, a disambiguating paragraph was constructed:

(20) a. *James and I were travelling in England last summer. An uncle of mine, the well-known Duke Charles of Northrup, invited us for dinner at his modest country home. When we arrived I introduced my uncle to James by saying, "Here is the famous Duke James." James was so impressed that he forgot to shake my uncle's hand!*

b. *While in England last summer I attended a ball at the Palace. All of royalty was there including some distinguished foreign guests. In particular, I noticed Duke James of Denmark and wanted to meet him. I was able to persuade my close friend Anna who speaks fluent Danish to introduce him to me. When they came over Anna said, "Here is the famous Duke James." I was overwhelmed by his charm.*

Ten speakers first familiarized themselves with a given paragraph by reading it silently to themselves. The speakers then read the paragraph aloud once for recording. The speakers were tested individually as in previous experiments, and the subjects who participated in this experiment were presented the paragraph contexts prior to the test reported previously with sentences in isolation, so as to avoid any awareness of the ambiguous nature of the sentences in this test or of their key status. In fact, subjects reported after reading the paragraphs that they were unaware of any ambiguity in the sentences contained therein. Other details of the procedure and acoustical analysis were identical to those of Experiment 3.2.3. The mean for the measurements of *P1*, *V*, and *P2* are given in Table 3.7 and are graphed in Figure 3.19.

Within-Sentence Comparisons

Comparison of *P1* vs. *V* revealed a significant fall in F_0 for the key word in both paragraph contexts ($p < .01$, t values of 3.39 and 5.21, $df = 9$). Comparison of *V* vs. *P2* showed a significant fall in F_0 from *V* to *P2* in the (a) reading ($p < .05$, $t = 2.52$, $df = 9$) and a nonsignificant rise in F_0 in the (b) reading ($p > .20$). Finally, comparison of *P1* vs. *P2* showed significantly lower value of *P2* in the (a) version ($p < .001$, $t = 5.74$, $df = 9$) and a nonsignificantly lower value of *P2* in the (b) version ($p > .20$).

Generally speaking, the results of the within-sentence comparisons closely parallel those obtained in the isolated sentences of Experiment 3.2.3 (compare

Table 3.7. Mean F_0 values for experiment 3.2.4

Paragraph		*P1*	*V*	*P2*
(20)	a.	190	162	146
	b.	166	160	164

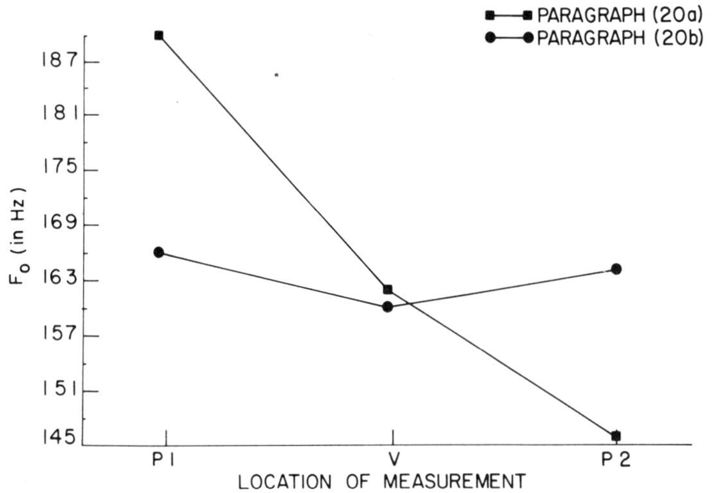

Figure 3.19. Plot of mean F_0 vs. time for the measured values of Paragraph Pair (20).

Figures 3.16 and 3.19). Most strikingly, the (a) version again exhibited a significant fall in F_0 from *V* to *P2*.

Cross-Sentence Comparisons

As in the previous experiment, the fall–rise patterns for the (a) and (b) versions differed in their overall shape. For the (a) version, a large amount of F_0 fall was exhibited across the boundary between the two key words, in addition to a fall within the first key word *Duke*. Also, the F_0 rise for the (a) version was contained within the second key word, unlike the pattern observed in the (b) version and in other cases of fall–rise patterns studied thus far. Again, the special status of the vocative in the (a) versions is implicated as a factor, independent of syntactic boundary strength.

A comparison of Figure 3.16 and 3.19 reveals that, while the general F_0 patterns were similar for these sentences when they were spoken in isolation and in paragraph contexts, the average values of F_0 were consistently lower for all measured points in the paragraph contexts. This difference may be attributable to the effect of paragraph declination, observed earlier by Lea (1973), among others, as discussed in Chapter 2. In effect, the absolute values of F_0 were probably lower in the paragraph context because the key sentences were spoken near the end of the paragraph declination function, whereas the same sentences spoken in isolation were not subject to this lowering function.

The results of this experiment generally provide support for the notion that the results for isolated sentence contexts generalize to paragraphs in which the speaker is unaware of the presence of an ambiguous string. Although future work will undoubtedly uncover other differences between F_0 contours in isolated sentences vs. paragraphs, it is noteworthy that the fall–rise patterns studied here seem relatively immune to this difference in context.

Conclusion

The results of this study indicate that the form and magnitude of fall–rise patterns can be influenced by syntactic boundary strength in a rather subtle manner. Here, the addition of a single phrase node of strength immediately preceding the key boundary typically produced an increment in the amount of F_0 fall. These results were in agreement with the theory of boundary strengths, in conjunction with the framework of phrase structure representation adopted at the outset of this chapter. As such, the data indicate that the strength theory is applicable to F_0 patterns and provide support for the idea that the type of phrase structure representation adopted on linguistic grounds provides a good first approximation to the type of internal syntactic representation computed by the speaker. At the same time, the results suggest that the falling and rising portions of fall–rise patterns are separably influenced by the left and right sides of a given syntactic boundary. In this sense, fall–rise patterns seem to differ from the other three prosodic attributes examined in terms of the theory of boundary strengths, these other attributes being unitary phenomena that seem to be influenced more by the combined strengths of both bordering phrases, as specified in Step 7 of the original algorithm (see Cooper and Paccia-Cooper, 1980).

Study 3.3. Deletion Sites[2]

In this study, fall–rise patterns were examined at phrase boundaries that coincided with the sites of syntactic deletion transformations. On intuitive grounds, it seemed that such locations are typically accompanied by rather pronounced fall–rise patterns in comparison to phrase boundaries that do not coincide with deletion sites, independent of the surface syntactic boundary strength. In earlier research, such deletion sites were found to be accompanied by segmental lengthening, pausing, and blocking of phonological rules. The magnitudes of these effects appeared greater than would be predicted on the basis of the syntactic surface boundaries alone. These findings led to the suggestion that speakers' internal syntactic representation might contain two separate levels, analogous to underlying and surface structures in a generative grammar. The speaker would have to possess knowledge about both levels of syntactic structure in order to process a deletion site, since a deletion site can only be specified with reference to an underlying level of syntactic structure in which the deleted material originally appeared. Because this line of reasoning carries such a major implication about the organization of the speaker's internal syntactic code, it seemed worthwhile to extend the study of deletion sites to fall–rise patterns of F_0. A preliminary examination of such patterns was provided by O'Shaughnessy (1976), although the results were largely inconsis-

[2]Part of this study appeared originally in Sorensen and Cooper (1980).

tent across the four speakers in that study. The experiments here focused on sites produced via the transformational rule of Gapping, which serves to delete material from within a major phrase (Jackendoff, 1971; Ross, 1970). Gapping was chosen for study because, unlike many other putative deletions, this rule receives relatively strong support from the linguistic literature as a bona fide deletion (Hankamer and Sag, 1976).

Experiment 3.3.1

This experiment focused on a type of Gapping known as Verb Gapping, in which a Verb is deleted under identity with a prior occurrence of the same word (Ross, 1970). Sentences were designed such that the same key words appeared at phrase boundaries in both deletion and nondeletion contexts. In each of eight sentence pairs, the (a) version contained no deletion at the key boundary, whereas the (b) version contained a deletion site produced by Verb Gapping. The eight sentence pairs appear below with the key words underlined.

(21) a. *The seamstress wove your hat and then made your scarf.*
b. *The seamstress wove your hat and the maid your scarf.*
(22) a. *The porter took your bags and weighed your luggage.*
b. *Roberta took your bags and Wade your luggage.*
(23) a. *The chef fixed the soup and then made your sandwich.*
b. *The chef fixed the soup and the maid your sandwich.*
(24) a. *The head teacher will help your daughter and then aid your son.*
b. *The head teacher will help your daughter and the aide your son.*
(25) a. *The bus driver will take your brother and then guide your sister.*
b. *The bus driver will take your brother and the guide your sister.*
(26) a. *The hostess will show your aunt and then Tom will guide your uncle.*
b. *The hostess will show your aunt and the touring guide your uncle.*
(27) a. *Aunt Bessie took your fruit and my uncle weighed your vegetables.*
b. *Aunt Bessie took your fruit and my uncle Wade your vegetables.*
(28) a. *The intern will help your father and the nurse will aid your mother.*
b. *The intern will help your father and the nurse's aide your mother.*

In each (b) sentence, the Verb has been deleted via Verb Gapping at the boundary between the key words. Thus, in (21b), for example, the Verb *wove* has been deleted under identity with the first occurrence of this word.

These sentences were originally designed for an experiment on the blocking of palatalization (Cooper, Egido, and Paccia, 1978), in which speakers were instructed to speak at a relatively fast rate in order to increase the general likelihood of palatalization across the key word boundary. In this experiment, however, a new group of speakers was given no special instructions regarding speech rate, since our interest here involved testing fall–rise patterns regardless of whether palatalization applied.

For a group of eight speakers, acoustical measurements of F_0 were obtained

Table 3.8. Mean F_0 values for experiment 3.3.1

Sentence		*P1*	*V*	*P2*
(21)	a.	191	182	184
	b.	184	171	176
(22)	a.	181	176	181
	b.	185	173	177
(23)	a.	193	179	184
	b.	188	173	180
(24)	a.	188	180	180
	b.	185	172	176
(25)	a.	185	173	177
	b.	184	166	176
(26)	a.	181	174	180
	b.	178	167	177
(27)	a.	188	178	181
	b.	191	171	174
(28)	a.	183	177	180
	b.	189	171	177
Grand-means	a.	186	177	181
	b.	186	171	177

for *P1* and *V* on the first key word of each sentence and for *P2* on the second key word. The mean values for these measurements are shown in Table 3.8. The mean values for the three measurements in the nongapped (a) and gapped (b) sentence versions are shown in Figure 3.20.

The results show fall–rise patterns in both deletion and nondeletion contexts, with the fall–rise pattern being more pronounced at the deletion sites. Furthermore, the comparison of *P1* and *P2* values indicates that the fall–rise patterns were local rather than a reflection of declination resetting. In particular, the average values of *P2* were lower than those of *P1* in 15 of the 16 test sentences, and the overall value of *P2* was 3.9% lower than that of *P1*.

Comparisons of the deletion and nondeletion contexts showed that the average fall from *P1* to *V* was almost twice as great in the deletion contexts, averaging 8.1% for the deletion sentences, vs. 4.5% for the nondeletion sentences. The average rise from *V* to *P2* was also consistently greater in the deletion contexts.

Comparison of *P1* vs. *V* showed significant falls in 15 of the 16 sentences ($p < .05$, t values ranging from 2.49 to 7.88, $df = 7$) and a nearly significant fall in Sentence (21b) ($p < .07$, $t = 2.19$, $df = 7$). Comparison of *V* vs. *P2* revealed significant rises in two of the (a) sentences [Sentence (26a), $p < .02$, $t = 3.10$, $df = 7$; Sentence (23a), $p < .05$, $t = 2.50$, $df = 7$] and in four of the (b) sentences, namely, Sentences (21b), (23b), (25b), and (26b) ($p < .05$, t values ranging from 2.56 to 4.98, $df = 7$). Nonsignificant rises from *V* to *P2* were observed in all remaining sentences ($p > .05$, t values ranging from 0.04 to 2.31, $df = 7$). Finally, comparison of *P1* vs. *P2* showed significantly higher values of *P1* in one of the (a) sentences [Sentence (27a), $p < .05$, $t = 2.86$, $df = 7$) and in three of the (b) sentences, namely, Sentences (22b), (27b), and (28b) ($p < .05$, t values ranging from 2.44 to 3.29, $df = 7$). These results suggest that the fall–rise patterns were superimposed on a single declination function, as expected at these within-clause sites (see Chapter 2; also O'Shaughnessy, 1976).

Comparisons of the amount of F_0 fall in the (a) vs. (b) sentences showed significantly greater fall in four of the (b) sentence versions ($p < .05$, t values ranging from 2.53 to 3.40, $df = 7$). Nonsignificant trends in the same direction were observed in the remaining four sentence pairs. As shown in Figure 3.20, the greater fall in the deletion-containing sentences is primarily attributable to a lowered value of *V*, with approximately equal values of *P1* in the gapped and nongapped sentences. Hence, the greater fall in the gapped sentences cannot be readily attributed to an effect of stress heightening, as discussed in Chapter 2.

Comparisons of the amount of F_0 rise in the (a) vs. (b) sentences showed no significant differences. The values of both *V* and *P2* were considerably lower in the gapped (b) sentences (see Figure 3.20), but the amount of rise from *V* to *P2* was approximately equal in the gapped and nongapped sentences.

In conclusion, the results of this experiment provide evidence for the notion that fall–rise patterns are more pronounced at deletion sites. This effect was particularly apparent for the F_0 fall, which averaged approximately twice as great in the deletion contexts. However, other data on deletions must be considered before one can verify the claim that the deletion per se is responsible

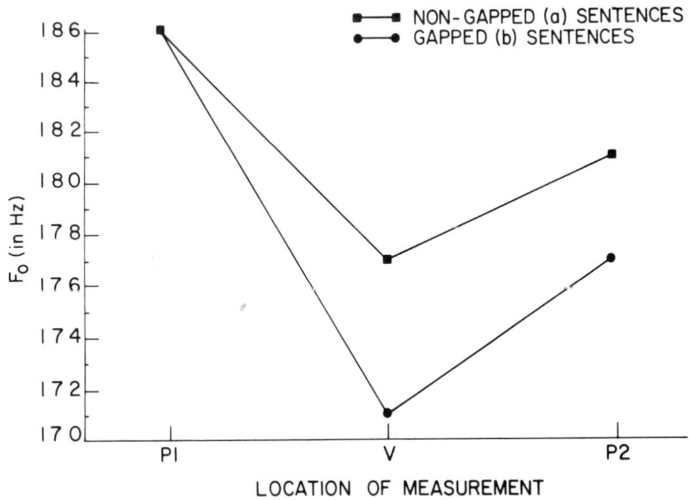

Figure 3.20. Plot of mean F_0 vs. time for the measured values pooled across Sentences (21)–(28) for the nongapped (a) and gapped (b) versions.

for the fall–rise in F_0, since the sentences produced by Verb Gapping also differ from nondeletion sentences in their surface structures. In particular, sentences with Verb Gapping exhibit a surface Noun Phrase–Noun Phrase sequence, whereas nondeletion sentences exhibit a Verb–Noun Phrase sequence. Although no independent evidence suggests that this particular surface structure might be responsible for the effects, it was still considered worthwhile to extend the study of deletion sites to another type of Gapping in which the same surface sequence was not present.

Experiment 3.3.2

To test the generality of the previous results, we conducted an experiment with a form of Gapping known as Noun Gapping. In this case, a Noun can be deleted under identity with a prior occurrence of the same Noun (Jackendoff, 1971). As in the previous experiment, sentence pairs were constructed with key words appearing at phrase boundaries in both deletion and nondeletion contexts. In each of two sentence pairs, the (a) version contained no deletion at the key boundary, whereas the (b) version contained a deletion site produced by Noun Gapping. The sentence pairs appear below with the key words underlined:

(29) a. *Janet had prune juice today before lunch and I had grape juice last week.*
b. *Janet had prune juice today before lunch and I had grape just last week.*

(30) a. *Janet made prune juice the day before yesterday and grape juice last week.*
b. *Janet made prune juice the day before yesterday and grape just last week.*

In each (b) sentence, the Noun of the direct object in the second clause has been deleted via Noun Gapping at the boundary between the key words. Thus, in (29b) for example, the Noun *juice* has been deleted under identity with the first occurrence of this word.

A group of eight speakers took part in this experiment, none of whom participated in the previous experiment. Acoustical measurements of F_0 were obtained for *P1* and *V* on the first key word of each sentence and for *P2* on the second key word. The mean values for these measurements appear in Table 3.9. Figure 3.21 shows the data for Pair (29), and Figure 3.22 for Pair (30).

Comparison of *P1* vs. *V* showed significant F_0 falls for each of the Gapped sentences [Sentence (29b), $p < .01$, $t = 5.27$, $df = 7$; Sentence (30b), $p < .01$, $t = 3.71$, $df = 7$] and nearly significant falls for each of the nongapped sentences [Sentence (29a), $p < .06$, $t = 2.34$, $df = 7$; Sentence (30a), $p < .06$, $t = 2.22$, $df = 7$]. The rise in F_0 from *V* to *P2* was significant in one of the gapped sentences [Sentence (29b), $p < .01$, $t = 3.45$, $df = 7$] and nearly

Table 3.9. Mean F_0 values for experiment 3.3.2

Sentence		*P1*	*V*	*P2*
(29)	a.	154	152	131
	b.	156	126	137
(30)	a.	161	159	135
	b.	150	125	136

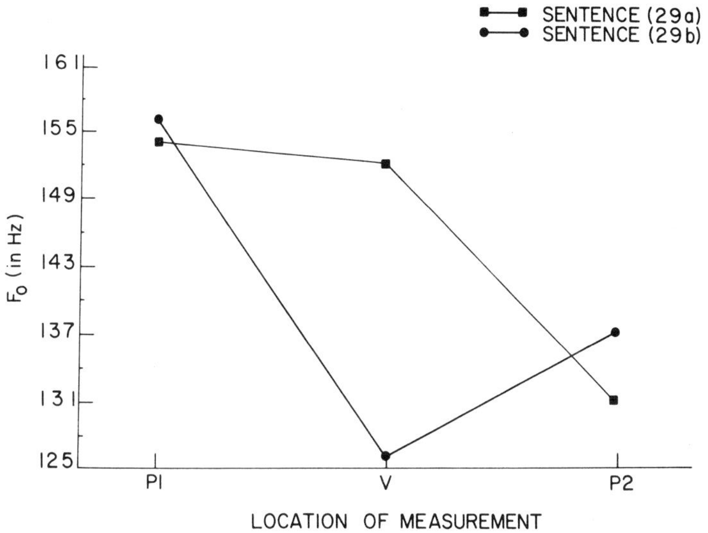

Figure 3.21. Plot of mean F_0 vs. time for the measured values of Sentence Pair (29).

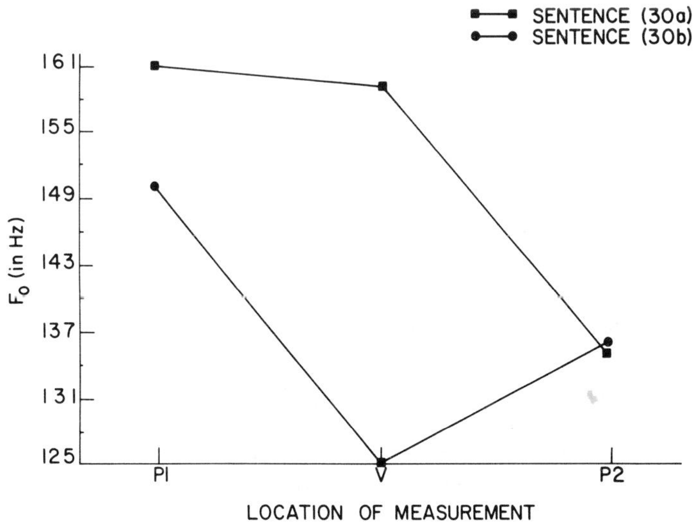

Figure 3.22. Plot of mean F_0 vs. time for the measured values of Sentence Pair (30).

significant in the other [Sentence (30b), $p < .06$, $t = 2.33$, $df = 7$]. In the nongapped sentences, there was actually a significant FALL from *V* to *P2* ($p < .001$, t values of 5.92 and 6.47, $df = 7$). Comparison of *P1* vs. *P2* showed significantly higher values of *P1* in each of the four sentences ($p < .01$, t values ranging from 3.77 to 7.30, $df = 7$). As in the case of Verb Gapping, it thus appears that the fall–rise patterns accompanying this deletion site were superimposed on a single declination function.

Comparison of the F_0 patterns in deleted and nondeleted sentences showed that the magnitude of both (*P1–V*) and (*V–P2*) was significantly greater in the sentences containing key deletion sites ($p < .01$, t values ranging from 3.55 to 5.90, $df = 7$). The fall–rise patterns for the sites of Noun Gapping were similar to those obtained at the sites of Verb Gapping in the previous experiment. The control sentences in this experiment did not exhibit fall–rise patterns at all, but instead showed a plateau + fall pattern, as indicated in Figures 3.21 and 3.22. This pattern typically accompanies an Adjective + Noun sequence such as the one contained at the key site in these sentences.

As in the case of Verb Gapping, the fall–rise patterns observed for Noun Gapping are primarily attributable to lowered values of *V* (see Figures 3.21 and 3.22). The values of *P1* were somewhat higher in (29a) than in (29b), and these values were about equal for the sentences in (30). Hence, the fall–rise patterns observed for Gapping again do not seem attributable to the stress heightening of the *P1* word. The values of *P2* were slightly higher in the gapped sentences, though not significantly.

From the results of this experiment alone, it is not possible to determine whether the substantially lower F_0 valley is attributable to the presence of a deletion site per se or to the fact that the first key word also appeared in constituent-final position in the surface structure of the gapped sentences only. To select between these two accounts, a follow-up experiment was conducted in which peak and valley values were compared for sentences in which a single key word always appeared in constituent-final position, as a noun–adjective homophone appearing (a) as the head noun at the end of a noun phrase in a sentence containing no deletion site vs. (b) as an adjective at the site of Noun Gapping, as in Sentences (31) and (32) below.

(31) a. *As for the dress supply, the manager of Marie's boutique ordered woolen dresses for the winter and the clerk ordered <u>silk</u> dresses for summer.*
     b. *As for the dress supply, the manager of Marie's boutique ordered woolen dresses during the winter and the clerk ordered <u>silk</u> during the summer.*

(32) a. *As for flowers, Angie and Raymond decided on white flowers for the hedge and <u>pink</u> flowers for the patio.*
     b. *As for flowers, Angie and Raymond decided on white flowers for the hedge and <u>pink</u> for the garden and patio.*

Since the key word appears in constituent-final position in both (a) and (b) sentences, any difference in the fall–rise patterns between these versions should be attributable to the presence of the deletion site. The results for these two sentence pairs with eight speakers showed a pattern similar to the results of the preceding experiment, with lower valley values for the (b) sentences containing the deletion site.[3] The valley values averaged 19 Hz lower in (31b) than (31a) and 13 Hz lower in (32b) than (32a); both effects were statistically significant ($p < .05$, t values of 2.48 and 4.13, $df = 7$).

Conclusion

The results of the experiments on both Verb and Noun Gapping indicate that these deletion sites are accompanied by significant fall–rise patterns in F_0. Such patterns are marked by a particularly large F_0 fall preceding the deletion site, the magnitude of this fall being greater than would be expected on the basis of the surface boundary strength at the location. This finding suggests that the F_0 fall is influenced by the presence of the deletion site per se. It is at least conceivable, however, that this exaggerated fall is attributable to a degree of awkwardness inevitably accompanying these locations. While we cannot rule out this possibility entirely, one piece of evidence against it involves the fact that the values of *P1* were approximately equal for the deletion and nondeletion contexts in these experiments. It seems more likely that the exaggerated F_0 fall that directly precedes a deletion site may be attributable to the speaker's processing of the deletion. In particular, the speaker might execute such an F_0 fall in order to signal the presence of the upcoming deletion site to the listener.

This line of reasoning carries with it the implication that the speaker's syntactic representation includes more than a single level of syntactic structure. In order for the speaker to program an F_0 effect based on information provided by a deletion site, the speaker must possess knowledge of an underlying level of syntactic structure from which such a deletion could have arisen. Thus, the present interpretation leads to the inference that the speaker's internal syntactic representation includes both underlying and surface levels of syntactic coding. Converging evidence in support of this inference is provided by the prosodic effects of pausing and blocking of cross-word phonetic conditioning (Cooper and Paccia-Cooper, 1980). Like the F_0 fall observed here, these additional prosodic effects are exaggerated at deletion sites in a manner that could not be readily predicted from a consideration of surface structure alone. While the combined results of these studies seem to implicate a non-surface syntactic influence on speech prosody, the results also indicate that this influence is both small in magnitude and applicable to a more restricted set of sentences than the influence of surface structure. Thus, we can maintain the conclusion that surface structure exerts the PRIMARY syntactic influence on the prosodic attributes under consideration here (Cooper and Paccia-Cooper, 1980) and that the secondary influence of non-surface syntax can be ignored for the purposes of many practical applications.

[3]The key words for these same utterances also exhibited longer durations at the deletion site (see Cooper and Paccia-Cooper, 1980).

General Conclusions

We have seen in this chapter that fall–rise patterns of F_0 accompany a variety of clause and phrase boundaries. The rank magnitude of the individual fall and rise components of such patterns varied systematically with the syntactic strength and location of the phrases bordering the boundary. Larger F_0 falls were produced for stronger phrases preceding the boundary, while larger F_0 rises were produced for stronger phrases following the boundary. In some cases, the addition of a single phrase node of boundary strength produced a significant increment in the magnitude of an F_0 movement, indicating that systematic variations in F_0 can be influenced by quite subtle variations in syntactic structure. In general, these results provided a source of converging evidence in favor of a type of syntactic representation in speech production that had previously been implicated as influencing other prosodic features, including segmental lengthening, pausing, and the blocking of cross-word phonetic conditioning effects. In each case, the rank magnitude of these effects at a given syntactic boundary varies directly with the strength of that boundary, as defined by the theory of boundary strengths.

For the most part, increases in the magnitude of fall–rise patterns were produced by a substantial lowering of the valley (*V*) and slightly higher values of the peak following the boundary (*P2*), with little or no variations in the values of the peak preceding the boundary (*P1*). These results suggest that the effects on fall–rise magnitude observed in these experiments appear to be under direct syntactic control and cannot be attributed to a local influence of stress heightening for the *P1* word.

While fall–rise patterns have been shown to accompany a variety of clause boundaries, phrase boundaries, and deletion sites, such patterns by no means accompany all boundaries between words. In Experiment 3.3.2, for example, the minor boundary between an Adjective and Noun was not accompanied by a fall–rise pattern, but rather by a plateau + fall pattern, commonly observed at minor boundaries (e.g., Maeda, 1976). Thus, a certain threshold of syntactic boundary strength is required to produce a fall–rise pattern. We cannot specify this threshold absolutely in syntactic terms, however, since its particular value is probably influenced somewhat by extrasyntactic factors as well, including speaking rate, constituent length, and characteristics of the individual speaker.

In particular, extrasyntactic factors seem attributable for the common observation that fall–rise patterns are often either small in magnitude or altogether absent at the boundary between NP and VP within a clause (e.g., Lea, 1973), despite the fact that such a boundary is relatively strong according to a metric of boundary strength. In many instances, the absence of significant fall–rise patterns at NP + VP boundaries may be attributed to the shortness of the NP subject constituent, often including only one word, as in *Tom saw the cat run away*. On the other hand, fall–rise patterns seem to be accentuated at boundaries between relatively long constituents. The effects of constituent length noted here are similar to those observed for segmental lengthening and pausing, and the length factor has already been incorporated into algorithms for determining the absolute magnitudes of these durational attributes (e.g., Bierwisch, 1966, Cooper and Paccia-Cooper, 1980; Grosjean, Grosjean, and Lane, 1979). However, more work needs to be conducted before the influences of constituent length on the magnitudes of fall–rise patterns is properly understood and quantified. A similar need exists for the study of a variety of other extrasyntactic factors.

The fall–rise patterns observed in this chapter have provided much evidence about the structural representation of the speaker's syntactic code, yet we still know virtually nothing about the processing operations that give rise to these patterns and their individual components. Perhaps the falling portion of F_0 in such patterns may be one by-product of a generalized relaxation response that accompanies the completion of a constituent. The generalized relaxation response would produce slackened vocal folds, yielding slower vocal fold movement, resulting ultimately in a lowered F_0. This same general response might, in addition, produce a slowed clocking of segmental timing, producing constituent-final lengthening. The rise in F_0 that accompanies the beginning of the next constituent could be accounted for, in principle, by a converse response of the processing machinery, by which the initiation of activity for a new constituent yields faster vocal fold movement and a faster clocking of segmental timing.[4] Although other factors may be superimposed to produce variations in F_0 and segmental timing that are at variance with this general framework, the present hypothesis does seem to provide a unified account of the direction of the F_0 and timing effects. In particular, this hypothesis provides a rationale for why F_0 patterns at syntactic boundaries are in the form of fall–rise rather than rise–fall and why segment durations are lengthened rather than shortened in phrase-final position.

[4]According to the present scheme, a word in phrase-initial position should be shortened relative to its duration in phrase-medial position, ceteris paribus. However, no such tendency has been observed for English (Cooper and Paccia, unpublished data, 1977), and phrase-initial words may actually be lengthened in the languages of Swedish (Lindblom and Rapp, 1973) and Italian (Nespor and Allen, 1977).

4. Blocking

In this chapter, we examine another kind of phenomenon in attempting to reveal how the speaker's syntactic representation influences aspects of F_0. In previous work, the concept of SYNTACTIC BLOCKING has been useful in describing constraints on phonetic conditioning effects that normally operate across word boundaries. According to this notion, conditioning effects may be blocked across major syntactic boundaries because the speaker does not have simultaneous access to phonetic information on both sides of the boundary, as required in order for such conditioning to apply (for a review, see Cooper and Paccia-Cooper, 1980). For example, consider the phonological rule of Palatalization, which converts an alveolar stop consonant (/d/ or /t/) into an affricate (/ǰ/ or /č/, respectively) in the environment of a following palatal (/y/). This rule operates to yield /koǰ(y)ɛt/ from *code yet* but is typically blocked when these two words span a major clause boundary, as in (1b) below:

(1) a. *We haven't broken the code yet but we intend to break it soon.*
    b. *We haven't broken the code yet we intend to break it soon.*

The effects of such blocking have been studied with a variety of conditioning rules, and the results of these studies point to the presence of syntactic structure as a direct influence on blocking, rather than as an indirect influence mediated by pausing (Cooper and Paccia-Cooper, 1980). Here, we attempted to investigate the possiblity of syntactic blocking for cross-word conditioning effects that involve F_0. At the outset, the effort was hampered by the fact that virtually no previous studies had examined conditioning effects for F_0 across word boundaries. In order to examine F_0 blocking, we first needed to establish a cross-word conditioning effect on F_0 that typically applied between words not separated by a major syntactic boundary.

The investigation included two studies: one involving a conditioning effect of stress; the other, a conditioning effect of voiced vs. voiceless stop consonants. In the first case, cross-word conditioning was first established between words not spanning major syntactic boundaries, and this effect was then successfully blocked when a major syntactic boundary intervened. In the second case, it was not possible to establish the very presence of a conditioning effect, but the experiments provide data relevant to Chapters 1 and 2 and illustrate some interesting problems of interpretation.

Study 4.1. Stress Blocking

Previous studies in English have shown that, when a word receives contrastive stress in a sentence, the F_0 of the stressed word is raised, whereas the F_0 of neighboring words is lowered (e.g., Atkinson, 1973). Since this latter effect represents a type of cross-word conditioning suitable for studying blocking, we designed experimental sentences to substantiate this F_0 lowering effect and to examine whether one could block the effect by the introduction of a sufficiently strong syntactic boundary. Nine subjects participated in this experiment, reading a list of six test sentences and two matched fillers. The test sentences appear below, with the key words underlined and contrastively stressed words capitalized:

(2) a. *In the summer I plan to bicycle across Canada with Pete and Jeanne while Freddy works on my car.*
  b. *In the summer I plan to bicycle across Canada with PETE and Jeanne while Freddy works on my car.*
  c. *In the summer I plan to bicycle across Canada with Pete and JEANNE while Freddy works on my car.*

(3) a. *In the summer I plan to bicycle across Canada with Pete and Jeanne will follow us in her car.*
  b. *In the summer I plan to bicycle across Canada with PETE and Jeanne will follow us in her car.*
  c. *In the summer I plan to bicycle across Canada with Pete and JEANNE will follow us in her car.*

The sentences of Group (2) can be characterized as follows:

(2′) a. $_S[\ldots X \ldots$ *Pete and Jeanne*$]_S\ _S[\ldots Y \ldots]_S$
  b. $_S[\ldots X \ldots$ *PETE and Jeanne*$]_S\ _S[\ldots Y \ldots]_S$
  c. $_S[\ldots X \ldots$ *Pete and JEANNE*$]_S\ _S[\ldots Y \ldots]_S$

The only difference among these three sentences involves the presence vs. absence and location of contrastive stress. The sentences thus provide a basic test of whether contrastive stress alters the F_0 values of neighboring words. In (2b), one can determine whether contrastive stress on *PETE* will alter the F_0 values on the following noun *Jeanne,* relative to (2a); and, in (2c), one can assess whether contrastive stress on *JEANNE* will likewise alter the F_0 values on the preceding noun *Pete,* relative to (2a).

Assuming that an effect of stress lowering is obtained for Group (2), the sentences of Group (3) provide a test of whether stress lowering can be blocked at major syntactic boundaries. In this Group, the key words *Pete* and *Jeanne* lie on opposite sides of a major clause boundary. The sentences of Group (3) can be schematized as follows:

(3′) a. $_S[\ldots X \ldots$ *Pete*$]_S$ $_S[$*and Jeanne* $\ldots Z \ldots]_S$
b. $_S[\ldots X \ldots$ *PETE*$]_S$ $_S[$*and Jeanne* $\ldots Z \ldots]_S$
c. $_S[\ldots X \ldots$ *Pete*$]_S$ $_S[$*and JEANNE* $\ldots Z \ldots]_S$

The material labeled as ... Y ... in (2′) and ... Z ... in (3′) is matched for number of syllables and overall stress pattern. The major difference between the two groups of sentences is that, in (2), the key words are contained in the same clause, whereas in (3), they appear in different clauses.

It is noteworthy that a control sentence, in which no contrastive stress appears, has been included in each of the two groups of sentences. With this design, it is possible to assess the effects of stress lowering for each Group in a manner that relies solely on the presence or absence of contrastive stress for sentences within that Group. As we shall see in Study 4.2, this important control cannot always be provided in instances of cross-word F_0 conditioning effects, making it difficult to attribute results to blocking or to accompanying effects imposed by the presence of a major clause boundary.

The subjects were instructed to place contrastive stress on any word which appeared in all capital letters. Four measurements were made in each sentence: the peak F_0 (*P1*) and valley F_0 (*V*) of the word *Pete*, the peak F_0 (*P2*) of *Jeanne*, and the duration of the silent interval from the offset of voicing for /i/ in *Pete* to the beginning of voicing for *and*. For a few utterances in which *Pete* was accompanied by a word-final release burst, the burst was included in this "silent" interval. Table 4.1 gives the mean and standard deviations for the measurements of *P1* and *P2*.

Let us first consider the data for the sentences of Group 2 to determine whether a cross-word conditioning effect was obtained. The mean peak F_0 for unstressed *Pete* and *Jeanne* in (2a) was 172 Hz and 162 Hz, respectively. Contrastive stress on *P1* in (2b) raised the mean peak F_0 of *Pete* from 172 Hz to 210 Hz, while the *P2* value of *Jeanne* was lowered from 162 Hz to 147 Hz. The difference between *P1* in (2a) and (2b) was significant [(2a) vs. (2b), $p < .001$, $t = -5.85$, $df = 8$] as was the difference between *P2* in the same sentences [(2a) vs. (2b), $p < .001$, $t = 6.30$, $df = 8$). In the absence of any other influence, we attribute this lowering of peak F_0 on *Jeanne* to the presence of contrastive stress on the word *Pete*.

Comparing (2a) and (2c), we note that contrastive stress on the *P2* word *Jeanne* raised the *P2* value significantly, from 162 Hz in (2a) to 186 Hz in (2c): [(2a) vs. (2c), $p < .01$, $t = -3.63$, $df = 8$]. However, stress on this word did NOT significantly lower the F_0 of the PRECEDING key word *Pete* [*P1* in (2a) vs. (*2c*), $p > .5$, $t = .44$, $df = 8$]. Apparently then, the effect of F_0 stress lowering only works in the forward direction, such that contrastive stress on a particular word acts to lower the F_0 value of a following word but not the

Table 4.1.[a] Means and standard deviations of F_0 values (Hz) for experiment 4.1.1

Sentence			Measurement *P1*	*P2*
(2)	a.	$\overline{F}_0$	172.	162.
		SD	48.1	42.1
	b.	$\overline{F}_0$	(210.)	147.
		SD	(51.7)	44.4
	c.	$\overline{F}_0$	171.	(186.)
		SD	49.7	(55.7)
(3)	a.	$\overline{F}_0$	172.	171.
		SD	47.2	49.1
	b.	$\overline{F}_0$	(211.)	173.
		SD	(53.4)	50.1
	c.	$\overline{F}_0$	167.	(196.)
		SD	43.6	(49.4)

[a]Measurements in parentheses indicate that the measured key word was spoken with contrastive stress.

preceding word. Thus, blocking can only be tested for the (a) vs. (b) sentences of Group (3).

Contrastive stress on *Pete* in (3b) raised the mean peak F_0 to 211 Hz, from 172 Hz in (3a). This difference was significant [(3a) vs. (3b), $p < .002$, $t = 4.56$, $df = 8$]. Furthermore, there was no F_0 lowering effect on *Jeanne* in (3b) relative to (3a); in fact, the mean peak F_0 was slightly higher in (3b) [(3b) vs (3a), $p > .2$, $t = .91$, $df = 8$]. It thus appears that the influence of F_0 lowering can be blocked by an intervening syntactic boundary.

Several differences exist in the type of cross-word phonological effect studied here compared with those studied previously (Cooper and Paccia-Cooper, 1980). In particular, with phonological rules such as Palatalization or Flapping, no pause appears between the two phonological segments when the rule operates. With the F_0 lowering effect noted in the Group (2) sentences, a pause appears between the stressed word and the word which is lowered in F_0. In Sentence (2b), the silent interval from *Pete* to *and* averaged 192 msec, with a standard deviation of 122 msec. Clearly then, the influence of F_0 stress lowering operates across a pause. In Sentence (3b), where we noted blocking of the F_0 stress lowering effect, the pause from *Pete* to *and* increased in duration to 445 msec, with a standard deviation of 187 msec. Thus, while pauses accompany both the application and blocking of the F_0 lowering effect, the duration of these pauses is typically longer when the effect is blocked.

Conclusion

The results of this study provide evidence for the blocking of a cross-word F_0 conditioning effect. In this case, the conditioning effect involves the lowering of F_0 on a word that follows a contrastively stressed word. The results show that this lowering effect can be blocked at the boundary between two major clauses; that is, when the contrastively stressed word appears at the end of one clause and the measured word appears at the beginning of the next clause. The blocking effect appears to reflect a constraint on the speaker's ability to "look" ahead and program prosodic information for an upcoming clause.

Study 4.2. Voiced–Voiceless Blocking[1]

Let us now consider another F_0 conditioning effect and test its operation across word boundaries. This effect, reviewed in Chapter 1, involves a fall in F_0 during the first 50 or so msec of a vowel conditioned by the presence of a prior voiceless consonant (e.g., /k/). Prior work (Lea, 1973; Stevens, Henke, and Sorensen, 1976) has shown that this effect operates within words, and we conducted a pilot study on two speakers to determine whether it also operates across word boundaries, i.e., when /k/ is word-final and the vowel is at the beginning of the next word. The results suggested the presence of a cross-word effect, although, as we shall see later, this result is subject to an alternative interpretation. After /k/, the F_0 for the word-initial vowel showed a falling contour during the first 50 msec. With this information, we were prepared to conduct experiments to test whether the cross-word effect could be blocked at major syntactic boundaries.

Experiment 4.2.1

In the first experiment, four pairs of sentences were constructed whose members contained a word boundary between a word ending in /k/ and a following word beginning in a vowel. The members of each pair differed according to whether the critical word boundary coincided with a major clause boundary or with a minor boundary between an Adjective and head Noun of a Noun Phrase.

The test sentences appear below with the key words underlined. In each pair, the (a) sentence contains the major syntactic break at the key word boundary.

[1]Portions of this study are revised from Sorensen and Cooper (1980), which appeared in R. A. Cole (ed.), *Perception and Production of Fluent Speech,* Lawrence Erlbaum Associates, Suite 112, 365 Broadway, Hillsdale, N.J. 07642. We thank Dr. Ronald A. Cole and the publisher for permission to adapt this material.

(4) a. *Unless they're black earrings won't look good with Jan's outfit.* (S-S)
b. *I said that black earrings would look good with Jan's outfit.* (Adj.-N)

(5) a. *Because they're sick eagles will be removed from the aviary.* (S-S)
b. *On Tuesday sick eagles will be removed from the aviary.* (Adj.-N)

(6) a. *Although I usually don't like my cereal thick oatmeal with brown sugar is good.* (S-S)
(I usually don't like my cereal to be thick)
b. *Although I usually don't care for cereal thick oatmeal with brown sugar is good.* (Adj.-N) (I like oatmeal)

(7) a. *Unless they're black otters are easy prey to sea lions.* (S-S)
b. *In Russia black otters are easy prey to sea lions.* (Adj.-N)

Ten speakers read these test sentences from a list interspersed with filler sentences. Acoustical measurements of F_0 were obtained at four successive locations: (1) the offset of voicing preceding the word-final /k/, (2) the onset of voicing for the word-initial vowel, (3) 50 msec after the onset of voicing for this vowel, and (4) 100 msec after the onset of voicing for this vowel.

The results for the ten speakers are given in Table 4.2 and summarized graphically in Figure 4.1. The (b) sentences exhibited a falling F_0 contour during the first 100 msec of the word-initial vowel, indicating the presence of a cross-word phonetic effect as in the pilot study. The (a) sentences, however, exhibited a sharply rising F_0 during the first 50 msec of the vowel. This sharp

Table 4.2. Means and standard deviations of F_0 values (Hz) for experiment 4.2.1

			Location of measurement			
Sentence			Voicing offset	Vowel onset	Onset + 50 msec	Onset + 100 msec
(4)	a.	$\overline{F}_0$	145.	162.	173.	178.
		SD	58.5	61.4	59.2	63.8
	b.	$\overline{F}_0$	174.	181.	179.	173.
		SD	58.4	62.7	59.6	60.0
(5)	a.	$\overline{F}_0$	143.	177.	180.	177.
		SD	48.1	58.9	58.5	55.6
	b.	$\overline{F}_0$	186.	182.	175.	165.
		SD	56.3	56.1	52.2	53.5
(6)	a.	$\overline{F}_0$	134.	161.	176.	169.
		SD	50.2	48.7	52.7	55.2
	b.	$\overline{F}_0$	194.	182.	182.	175.
		SD	69.2	57.2	61.3	56.5
(7)	a.	$\overline{F}_0$	143.	151.	166.	173.
		SD	53.6	49.4	53.4	53.8
	b.	$\overline{F}_0$	180.	169.	167.	166.
		SD	61.7	54.3	54.9	59.5
Grand-means	a.	$\overline{F}_0$	142	163	174	175
	b.	$\overline{F}_0$	184	179	176	170

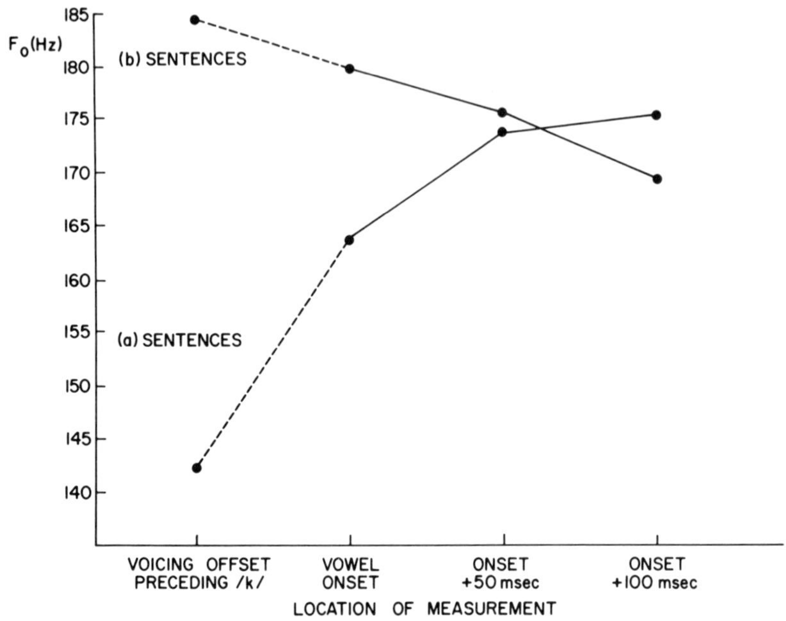

Figure 4.1. Plot of mean F_0 vs. time for each of the measured values in the (a) and (b) versions of Sentences (4)–(7), pooled across sentences. A dotted line connecting the first and second points of each curve represents the fact that these points were separated by a word boundary.

rise appears to reflect some type of start-up effect in F_0 at the beginning of a new clause. Since this effect was quite large, there was no way to tell whether the cross-word F_0 effect from the word-final /k/ had been blocked in the (a) sentences or simply overridden by a larger start-up effect in the opposite direction.

Experiment 4.2.2

Accordingly, we designed a second experiment in which further gradations of syntactic boundary strength were included, in the hope that a clause-internal phrase boundary rather than a major clause boundary would reveal an effect of blocking without producing a confounding start-up effect of F_0 rise. The four test sentences shown below were spoken by the same ten speakers as in the previous experiment. The (a)–(d) sentences range from strong to weak syntactic boundaries.

(8) a. *Unless you're coming back Allen will park in Terry's driveway.* (S-S)
 b. *I told the man in back Allen's new car was in the driveway.* (NP-Rel)
 c. *Diane will try to back Allen's new car into the driveway.* (V-NP)
 d. *I told Diane the back alley was made into a driveway.* (Adj.-N)

Table 4.3. Means and standard deviations of F_0 values (Hz) in experiment 4.2.2

Sentence			Location of measurement			
			Voicing offset	Vowel onset	Onset + 50 msec	Onset + 100 msec
(8)	a.	$\overline{F}_0$	138.	148.	157.	165.
		SD	49.5	38.4	44.2	48.0
	b.	$\overline{F}_0$	140.	146.	157.	160.
		SD	52.8	47.9	50.3	51.9
	c.	$\overline{F}_0$	153.	155.	157.	160.
		SD	51.4	65.5	53.5	50.9
	d.	$\overline{F}_0$	163.	165.	158.	154.
		SD	53.2	54.8	47.4	47.2

Acoustical measurements were obtained for the same four locations, and the results are given in Table 4.3. As in the previous experiment, a significant start-up F_0 rise during the vowel's first 50 msec was obtained for both the (a) and (b) sentences. For the Verb–Noun Phrase boundary in (c), however, the F_0 was nearly flat for the vowel, as expected if blocking occurred in the absence of any other F_0 effects. A slightly rising F_0 was observed for this vowel, but the conditioning effect of the preceding /k/ would have produced a falling contour.

In contrast, the (d) version with the minor Adjective–Noun boundary exhibited the falling F_0 contour indicative of an F_0 conditioning effect produced by the word-final /k/. The difference between the (c) and (d) sentences, shown graphically in Figure 4.2, suggested that the F_0 effect was indeed blocked in (c).

Experiment 4.2.3

A third experiment was conducted to focus exclusively on the difference between the Adjective–Noun and Verb–Noun Phrase boundaries. Ten new speakers uttered two pairs of sentences in which a key word boundary coincided with the contrast between the minor Adjective–Noun boundary (b) and the more major Verb–Noun Phrase boundary (a).

(9) a. *We found out that Joey will make eggs and hot muffins for his brother's lunch.* (V-NP)
b. *We found out that Joey used fake eggs and live rabbits in the magic show.* (Adj.-N)

(10) a. *Janice discovered that the Baxters will make ice in their freezer for ninety people.* (V-NP)
b. *Janice discovered that the Bruins use fake ice for their hockey games in the Garden.* (Adj.-N)

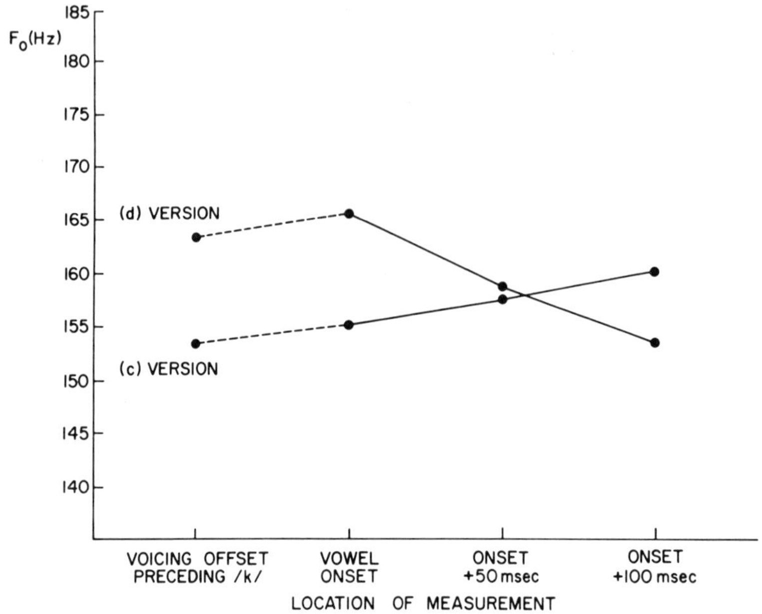

Figure 4.2. Plot of mean F_0 vs. time for each of the measured values in the (c) and (d) versions of Sentences (4)–(7), pooled across sentences. A dotted line connecting the first and second points of each curve represents the fact that these points were separated by a word boundary.

The results are given in Table 4.4 and graphed in Figure 4.3. As in the previous experiment, the F_0 contour during the first 100 msec of the word-initial vowel was virtually flat for the Verb–Noun Phrase boundaries of the (a) sentences, whereas the F_0 contour showed a falling pattern for the Adjective–Noun boundaries in (b). Again, it appeared that the F_0 contour of the vowel was influenced by the presence of a preceding word-final /k/ in the minor Adjective–Noun boundary but was not influenced by the presence of this same segment in the major Verb–Noun Phrase boundary.

It is tempting to conclude that the flat F_0 contour for the Verb–Noun Phrase boundary represents the blocking of the /k/ conditioning effect. However, it is also possible that the differences between the minor and major boundaries can be attributed to differences in inherent stress. With an Adjective–Noun sequence, speakers sometimes produce a rising F_0 on the Adjective and a falling F_0 on the Noun (e.g., Maeda, 1976; O'Shaughnessy, 1976). Thus, it is not clear on the basis of these data alone that the /k/ actually exerts a cross-word effect in the Adjective–Noun case. What is needed is an independent source of evidence to verify whether /k/ produces a cross-word phonetic effect on the F_0 contour. If no cross-word effect is actually present following /k/, then of course there is nothing to block by introducing a stronger boundary. Since the boundary between a Verb and Noun Phrase is typically flat in F_0, the results obtained

Table 4.4. Means and standard deviations of F_0 values (Hz) for experiment 4.2.3

Sentence			Location of measurement			
			Voicing offset	Vowel onset	Onset + 50 msec	Onset + 100 msec
(9)	a.	$\bar{F}_0$	112	120	120	120
		SD	28	39	27	30
	b.	$\bar{F}_0$	123	124	117	115
		SD	25	28	24	23
(10)	a.	$\bar{F}_0$	114	119	120	119
		SD	25	29	26	24
	b.	$\bar{F}_0$	120	118	116	113
		SD	28	32	29	28

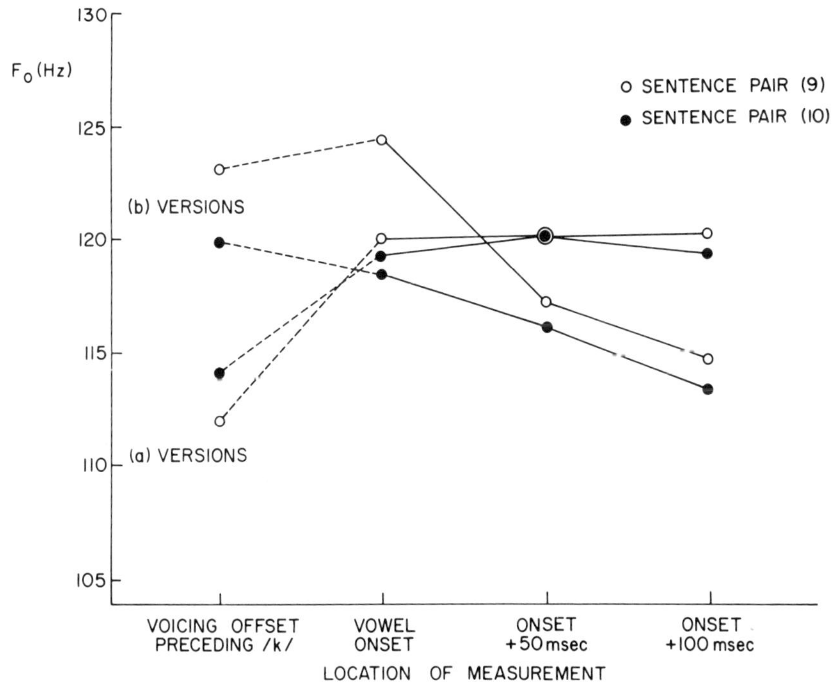

Figure 4.3. Plot of mean F_0 vs. time for each of the measured values in Sentence Pairs (9) and (10). A dotted line connecting the first and second points of each curve represents the fact that these points were separated by a word boundary.

in this experiment could be attributed solely to inherent differences in F_0 as a function of phrase structure.

Experiment 4.2.4

Given the results of the first three experiments of this study, we were still faced with the following question: Is the gradually falling F_0 contour on the first 100 msec of the vowel at the Adjective–Noun boundaries of the (b) sentences of Experiment 4.2.1, sentence (d) in Experiment 4.2.2, and the (b) sentences of Experiment 4.2.3 due to a cross-word phonetic effect of F_0 or to inherent stress pattern? To answer this question, we examined a converse F_0 conditioning effect of the *voiced* stop consonant /g/. Previous work indicated that, in within-word contexts, /g/ produces an F_0 RISE during the following vowel, whereas, as already shown, /k/ produces an F_0 fall on the following vowel (Lea, 1973, pp. 17–70). An experiment was designed to examine whether word-final /g/ shows any cross-word F_0 effect. Such an effect would be revealed as a gradually rising F_0 contour during the first 100 msec of an initial vowel. Since the inherent stress account predicts a gradually falling F_0 at Adjective–Noun boundaries, and the phonetic effect of /g/ predicts a rise in F_0, the results should permit us to distinguish between these two alternatives.

Ten new speakers uttered the following two pairs of test sentences.

(11) a. *Sandy asked George to <u>dig</u> <u>onions</u> out of the basket on the porch.* (V-NP)
b. *Sandy tossed all the <u>big</u> <u>onions</u> into the basket on the porch.* (Adj.-N)

(12) a. *Sandy asked George to <u>dig</u> <u>apples</u> out of the basket on the porch.* (V-NP)
b. *Sandy tossed all the <u>big</u> <u>apples</u> into the basket on the porch.* (Adj.-N)

Table 4.5. Means and standard deviations of F_0 values (Hz) for experiment 4.2.4

Sentence			Location of measurement			
			Voicing offset	Vowel onset	Onset + 50 msec	Onset + 100 msec
(11)	a.	$\bar{F}_0$	145	138	149	148
		SD	36	44	33	29
	b.	$\bar{F}_0$	154	144	148	148
		SD	37	38	30	30
(12)	a.	$\bar{F}_0$	138	137	145	145
		SD	34	45	34	30
	b.	$\bar{F}_0$	157	150	154	149
		SD	37	44	32	32

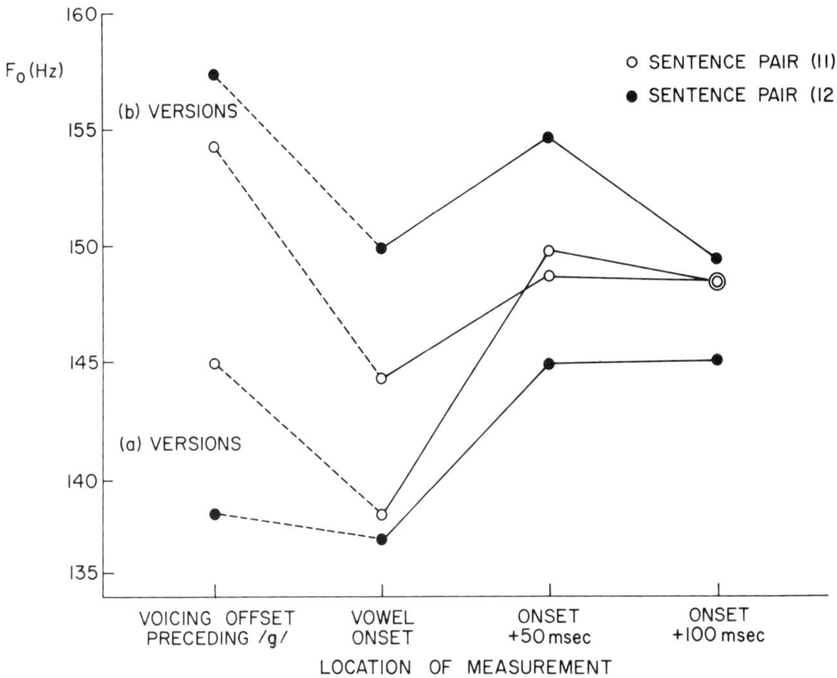

Figure 4.4. Plot of mean F_0 vs. time for each of the measured values in Sentence Pairs (11) and (12). A dotted line connecting the first and second points of each curve represents the fact that these points were separated by a word boundary.

The F_0 contours were measured as in previous experiments, and the data are presented in Table 4.5 and shown graphically in Figure 4.4. Figure 4.4 indicates a rise in F_0 for both (b) versions during the first 50 msec of the vowel. The finding is consistent with the concept that the consonants DO produce a cross-word effect on the F_0 contour.

The blocking hypothesis, however, predicts that the F_0 contours in the (a) versions should be flat. The data do not support this prediction. In fact, the rise in F_0 for the (a) versions was greater in magnitude than for the (b) versions. We can conclude that any cross-word phonetic effect of /g/ was not blocked at the Verb–Noun Phrase boundary. This conclusion is not surprising because most subjects exhibited continuous voicing across both the Adjective–Noun and Verb–Noun Phrase boundaries. In work with other cross-word phonetic effects, it has been shown that the presence or absence of blocking at syntactic boundaries does not correlate with the MAGNITUDE of pause duration (Cooper and Paccia-Cooper, 1980). However, when the speakers produce no pause whatsoever, as with most of the present utterances, it seems likely that the speakers could simultaneously access phonetic information on both sides of the boundary.

In order to examine blocking in this phonetic environment, it would thus seem necessary to extend the study to stronger syntactic boundaries in which continuous voicing might not appear. However, stronger syntactic boundaries are also accompanied by independent effects that would mask any true effects of blocking, as in Experiment 4.2.1. Accordingly, we did not attempt to insert a larger boundary between the /g/-final and vowel-initial words.

The results of this experiment with /g/ do cast a shadow on the previous results with /k/ at these boundaries, suggesting that blocking, in the sense described earlier, might not have occurred with /k/ at the Verb–Noun Phrase boundary. While it is possible that blocking did occur with /k/ but did not occur with /g/, due to the continuous voicing across the word boundary in the latter case, the differences noted for /k/ may have been due to stress influences other than blocking, as noted earlier. This problem of interpretation notwithstanding, a few pieces of information have been obtained in this study about the substantial effects of clause boundaries on F_0 inflections. In addition, the exercise has pointed out the importance of the design features that could be implemented in the stress lowering experiment in Study 4.1, in which it was possible to provide a control for overriding additional effects that accompany constituent boundaries. Thanks to this control capability, it was possible to ascertain the clear presence of blocking in that experiment.

General Conclusions

In this chapter, we have studied the syntactic blocking of cross-word conditioning effects on F_0 in two different situations. In Study 4.1, a conditioning effect of F_0 lowering was provided by the placement of contrastive stress on a preceding word. The experimental results clearly indicated that this effect operated across word boundaries when the key words were not separated by a major syntactic boundary. However, when a major clause boundary intervened, the effect of F_0 lowering was systematically blocked, suggesting that the speaker did not have simultaneous access to both words at the stage of coding at which the F_0 lowering effect normally applies. In Study 4.2, the conditioning effect of F_0 at the beginning of a word-initial vowel was intended to be provided by the placement of a voiced vs. voiceless stop consonant at the end of the preceding word. Although this particular conditioning effect had been demonstrated to operate reliably within words, experimentation failed to show clear evidence of an independent cross-word effect. However, the results of these experiments did provide additional information about the large-magnitude rise in F_0 that accompanies the beginning of a new clause, observed also in Chapter 2.

The blocking effect observed in Study 4.1 suggests that the speaker's normal access to more than a single word for programming prosodic effects is constrained by syntactic domains. Precisely why this kind of programming is so constrained remains a question for future research. One possibility is that blocking occurs near syntactic boundaries because the speaker must at this time devote additional processing to programming phrase-final prosodic effects and/or planning of the next phrase.

Another question that remains largely unexplored involves determining the specific strength of syntactic boundary that is sufficient to trigger blocking. As observed in studies on the blocking of other cross-word phonetic effects (Cooper and Paccia-Cooper, 1980), blocking of F_0 effects is expected to be partially determined by extrasyntactic factors as well, such as speaking rate and possibly constituent length. Apparently, the speaker can enlarge the domain for coding prosodic information in cases of fast rates of speech, in particular, suggesting that a stronger syntactic boundary would be required to trigger blocking at fast rates than at normal rates. Because the speaker's coding domains are influenced by both syntactic and extrasyntactic factors, it will be a considerable challenge to develop a theory of blocking that is applicable to a typical range of speaking circumstances.

5. Conclusions

Consolidation

Of the three phenomena examined in this study—declination, fall–rise patterns and blocking—all share a commonality in their capacity to be influenced by constraints on the speaker's domains of syntactic coding. In the case of declination, it was observed in Chapter 2 that the topline function typically resets to a new high starting value at a major boundary between clauses of at least moderate length. Moreover, an analysis of pauses at the major boundary suggested that this resetting was triggered by the presence of the syntactic boundary rather than by an intermediary influence of a breath pause. The speaker's domain of programming for declination thus seems to be defined, in part, in terms of a major syntactic constituent. In short utterances, this domain encompasses the entire sentence; in longer utterances, it typically corresponds to the main clause. While the domain of application for declination is largely dependent on the speaker's syntactic coding, other factors, including speaking rate and constituent length, also probably exert some influence.

In the case of fall–rise patterns of F_0, syntactic domains figure prominently in any account of the locations at which such patterns are observed in a sentence. The experiments of Chapter 3 indicated that fall–rise patterns accompany a large variety of both clause and phrase boundaries, whereas such patterns do not typically appear between words separated by only very minor syntactic boundaries. Because fall–rise patterns accompany a large range of syntactic boundaries, unlike declination resetting, such patterns offered an opportunity to test fine gradations of the theory of syntactic boundary strengths developed by Cooper and Paccia-Cooper (1980). The results showed that the rank magnitude of the falling and rising portions of the fall–rise patterns in different sentences could be predicted according to the syntactic strength of the phrases immediately preceding and following the boundary location, respectively. In some cases, a significant increment in the magnitude of the falling or rising portion of a fall–rise pattern was obtained by the addition of a single phrase node in the syntactic representation of the sentence, suggesting that speakers program rather fine-grained syntactic structures during speech production. Indeed, the systematic study of increments in the magnitude of fall–

rise patterns seems to reveal effects of such sensitivity that it should continue to provide a tool for testing hypotheses about whether a speaker's syntactic representation contains phrase structure nodes whose existence may be controversial on purely linguistic grounds.

Whereas declination represents a global feature of F_0 that seems to be programmed over a domain encompassing a major syntactic constituent, fall–rise patterns appear locally at the sites of a variety of syntactic boundaries and are superimposed on the more global declination function. Fall–rise patterns are constrained by the speaker's domains of syntactic coding in that such patterns appear primarily at syntactic boundaries. The falling portion of the F_0 pattern appears to reflect a relaxation response of the speaker upon the completion of executing one phrasal constituent, while the rising portion of the pattern seems to reflect a start-up effect upon the initiation of the next constituent. The domains of constituent coding that trigger fall–rise patterns include both clauses and subordinate phrases, whereas the domains that trigger declination resetting are largely restricted to main clauses or larger units. The blocking of cross-word F_0 conditioning is also influenced by the speaker's syntactic domains. In the study of stress blocking in Chapter 4, blocking was observed at the boundary between two main clauses but not between words separated by only a minor syntactic boundary. The blocking effect can be accounted for by assuming that the speaker's domain for applying cross-word conditioning effects normally encompasses two or more words but is prohibited from look-ahead in cases where the next word spans a major syntactic boundary. Thus, the speaker's programming of cross-word conditioning effects is constrained by the domain of syntactic coding in much the same way as declination. So far, blocking has only been established for F_0 conditioning in the case of major clause boundaries, and it remains to be determined whether such blocking accompanies some subordinate phrase boundaries as well.

In summary, the notion of syntactic domain appears to provide a general account of the locations in an utterance at which the three F_0 phenomena examined in this study are most likely to occur. Not only does this notion account for the most likely locations of these three phenomena, it also leads, when coupled with the theory of syntactic boundary strengths, to an account of the rank magnitudes of the falling and rising portions of fall–rise patterns, the one type of F_0 phenomenon in which variations in magnitude have thus far been examined.

While the theory of syntactic boundary strengths provides a first-order account of the locations of each of the three F_0 phenomena as well as an account of the rank magnitudes of the individual components of fall–rise patterns, it does not, by itself, explain the direction of the F_0 excursions exhibited by these phenomena. That is, a consideration of syntactic structures per se does not provide an explanation for why, in the case of declination, F_0 gradually falls rather than rises throughout the course of a main clause or why, in the case of fall–rise patterns, speakers produce the sequence fall–rise rather than rise–fall. In the latter case, a first-order account of the direction of the effect was proposed in Chapter 3 in terms of a possible sequence of general relaxation and tensing responses of the speaker upon the ending and initiation of phrases, respectively. In the case of declination, the gradual falling pattern of F_0 throughout the course of a main clause may be attributed to the combined weight of a number of factors, including a gradual decrease in subglottal pressure and a gradual slackening of the speech musculature, including the vocal folds. It is important to recognize that, while the direction of declination can be accounted for in terms of a physiological "running down" of the speech system, this general tendency can be outweighed by independent factors in any particular utterance, and, in at least one Brazilian language, declination is not observed regularly at all in declarative speech, as mentioned in Chapter 1. Thus, declination is by no means a physiological necessity.

Aside from a general treatment of the location, rank magnitude, and direction of F_0 phenomena, the results of this study point up the need to develop an account of the precise form exhibited by the topline function of declination. In Chapter 2, the Topline Rule provided a mathematical means of predicting the F_0 values of intermediate F_0 peaks in a sentence given the F_0 values of the first and last peaks of the sentence as well as their associated times of occurrence. The Topline Rule proved quite powerful in its ability to predict the average peak F_0 values in a large variety of sentence types. Because this rule can be applied invariantly across such a wide range of sentences and can also predict with reasonable accuracy the peak F_0 values in sentences spoken at different rates and by different speakers, it appears to capture much of the essence of F_0 declination in the declarative speech of American English, at least during oral reading as examined here. Yet, it is not possible to provide an explanation for the Topline Rule in terms of the physiological underpinnings of the speech system. Although a major component of the topline function is probably attributable to the combined weight of a decline in subglottal pressure and slackened vocal folds (e.g., Collier, 1975; Maeda, 1976), additional factors, some of which are language dependent, also play a role, as suggested by the inability to apply the Topline Rule to Japanese in Chapter 2.

Although much remains to be understood about the nature of the Topline Rule, the discovery of the rule itself may be the single most significant outcome of the empirical work of this study. Not only does the rule capture a form of declination that is largely independent of wide variations in sentence type, it also provides the first real basis for factoring out this "dc" component in future quantitative studies of local F_0 inflections that are superimposed on the declination function.

New Directions

The present study has provided a basis for much further testing of F_0 contours in sentences and larger contexts. The results of the experiments have demonstrated that a variety of issues can be tested with a sentence-reading procedure in which a large number of speakers utter sentences specifically designed to permit systematic hypothesis-testing and the application of statistical analyses. For the most part, this degree of experimental control was not brought to bear on the study of F_0 in previous research, under the rationale that the great complexity of F_0 contours could not be subjected to hypothesis testing at this early stage in our knowledge. Yet, the results of this study seem to indicate that rigorous experimentation can be applied to this topic without trivializing the issues or ignoring at least some aspects of the complexity of F_0 contours themselves. At the same time, it is clear that the examination of F_0 patterns in spontaneous speech allows one to consider many important issues that cannot be addressed within the present framework, and only future research can determine the extent to which our results are useful in guiding efforts to apply hypothesis testing to such a naturalistic setting. We suspect that studies of spontaneous speech will show points of both convergence and divergence with the data obtained for oral sentence-reading, and, by having the latter as a basis for comparison, it should be possible to determine which new findings are unique to spontaneous activity.

But it also seems that much useful work remains to be conducted within the confines of the controlled experimental procedure itself, including tests designed to assess the generality of the present findings over a broader range of sentences. In addition, studies might be conducted to examine a variety of extrasyntactic variables and their possible influence on the three F_0 phenomena tested here. These extrasyntactic variables include constituent length, speaking rate (examined with respect to declination in Chapter 2 but not yet tested with fall–rise patterns or blocking phenomena), semantics, and pragmatics, among others.

In each case, it seems possible that the choice of syntactic domains of coding by the speaker can be influenced in part by any of these factors. For example, the speaker may select a domain of coding that is more superordinate than normal when the constituents contained in this domain are short in terms of their number of words or syllables, when speaking fast, and when the material contained in the domain is highly cohesive on semantic and/or pragmatic grounds. If we have been correct in assuming that a speaker's syntactic domains of coding play a determining role in triggering each of the major phenomena studied here, then the influence of these extrasyntactic factors should likewise be revealed by examining any of these phenomena. In the present case, it would be predicted that the speaker's selection of a more superordinate syntactic domain would result in a larger-than-normal domain for declination, with resetting typically accompanying only the boundary between two sentences, fewer fall–rise patterns at relatively minor syntactic boundaries, and less blocking of cross-word conditioning at all but the most major syntactic boundaries. Converse effects would be predicted for the situation in which the speaker selects a domain of syntactic coding that is more subordinate than normal, triggered by longer constituents, slower speech, and/or lack of semantic or pragmatic cohesion. If these predictions are confirmed in future work, then it would appear that the three major F_0 phenomena examined here are subject to highly similar constraints.

In some cases, it is conceivable that influences on F_0 which have been characterized syntactically in this study might eventually be recast in semantic or pragmatic terms. For the most part, however, the use of controlled sentence materials in this study has allowed us to examine many of the syntactic influences on F_0 in situations that isolate the syntactic component as nearly as the English language permits. This focus on syntax was adopted largely because a number of explicit hypotheses that bear implications for the speaker's complex internal code could be tested on this topic. Nonetheless, the emphasis placed on syntax here should not be taken as an indication that extrasyntactic influences on F_0 are ultimately less worthy of study in their own right. On the contrary, some extrasyntactic factors, as suggested above, probably play a quite prominent role in influencing the form and magnitude of many F_0 effects, and the development of a theory that can adequately integrate the syntactic and extrasyntactic factors represents a major goal of ongoing research.

The present method of testing might also be applied to additional F_0 characteristics. A number of F_0 attributes have been documented in previous analyses (e.g., Maeda, 1976), and it seems likely that we are just beginning to tap the richness of F_0 as a source of information about the speaker's internal coding of syntactic and extrasyntactic variables.

On a broader scale, it might be helpful to examine the possible interrelationships involving the speaker's programming of different prosodic attributes that seem to be influenced in a similar manner by the syntactic code. By adopting a highly analytic approach in this study, it has been possible to characterize some of the more salient aspects of F_0 patterns. As noted earlier, similar investigations have been carried out with other prosodic attributes as well, and the stage now seems set for an effort to begin studying how these various features are integrated as part of the speaker's overall production system. An understanding of such integration is important both theoretically and for practical applications such as speech synthesis-by-rule, to be discussed in Chapter 6. In the case of F_0 and segmental timing, some preliminary research has already been directed to this general issue. In a recent study of Swedish speech, Lyberg (1979) has suggested that phrase-final lengthening is produced as a consequence of the speaker's need to execute a phrase-final fall in F_0 (Bolinger, 1964). Based on data for two speakers, Lyberg argued that a causal relationship exists between F_0 and phrase-final lengthening because the duration of a phrase-final stressed syllable was found to be longer as a monosyllabic word than when it appeared as the first syllable of a phrase-final bisyllabic trochee. It was suggested that the difference in these durations was attributed to the fact that the F_0 fall was produced entirely within the stressed monosyllabic word, whereas much of this fall could be produced on the second, unstressed syllable of the bysyllabic word, avoiding the need for lengthening the first, stressed syllable of the same word. There seems to be reason to doubt, however, that these data warrant conclusion about a causal relationship between F_0 and segmental lengthening. It is possible that the stressed syllable was lengthened to a greater extent as a monosyllable in absolute phrase-final position simply because of its closer proximity to the phrase boundary and/or to its status as a monosyllable, without invoking F_0 as a determining factor.

To test whether phrase-final lengthening and F_0 fall patterns are interrelated, Cooper and Sorensen (1977) examined the magnitude of lengthening and F_0 falls in individual utterances of Study 3.1.3. The results showed no significant correlation between the magnitudes of the phrase-final effects for timing and F_0, suggesting that these two attributes might be programmed independent of one another to a greater degree than was previously expected. However, a good deal more work remains to be conducted on the possible relationships between timing and F_0. At a general level, it is certain that the speaker's programming of a larger F_0 inflection is contingent upon the programming of a sufficiently long time interval during which to execute such an inflection (as discussed, for example, in Experiment 3.1.1). It is equally clear that the speaker's programming of F_0 and timing are influenced in a highly similar manner by syntactic structure; i.e., the magnitudes of fall–rise patterns and segmental lengthening are generally larger at stronger syntactic boundaries, ceteris paribus. What remains quite unclear is whether, at a more microscopic level, the speaker's programming of F_0 and timing are intimately related during the production of any individual utterance.

6. Extensions

In attempting to apply rigorous experimental techniques to the study of F_0 patterns, we have of necessity restricted the scope of this investigation. However, the realm of F_0 encompasses a much larger array of issues than has been examined experimentally in Chapters 2–4 and discussed in Chapter 5. In this chapter, we explore some of these areas, to present what is known already and to propose some new ideas as well, based on the findings of this study. For the most part, investigations of these related issues have so far not included the kind of precise quantitative measures presented in this study, but they nevertheless provide a number of helpful clues about topics to which future quantitative studies can be directed. In some cases, as discussed below, the methods and results of this study bear particular implications for research on issues involving F_0 which, at another level of analysis, seem to be quite remote from the considerations discussed in earlier chapters.

Language Development (From Infancy through Adulthood)

In the present research, the scope of investigation was restricted to young adult speakers ranging in age from about 18 to 30 years of age. However, the characteristics of F_0 for a given speaker undergo systematic changes during the speaker's lifetime, and developmental studies of F_0 can provide information on a variety of issues. During infancy and childhood, these changes may be attributed to maturational development of the vocal apparatus as well as to the development of cognitive and linguistic skills. Environmental factors play an increasingly important role in changing the characteristics of F_0 throughout childhood and adulthood.

During the first days of infancy, a speaker produces some of the highest F_0 values of his or her entire lifetime. The cries of infancy contain F_0 values as high as 400–600 Hz (Lind, 1965; Murray, Amundson, and Hollien, 1977; Wasz-Hockert, Lind, Vuorenkoski, Partanen, and Valanne, 1968; Wolff, 1969). Acoustical analysis of crying has proved useful as a diagnostic tool for detecting abnormalities in young infants (e.g., Wasz-Hockert et al., 1968;

Ostwald, 1972; Zeskind and Lester, 1978; Colton and Steinschneider, 1979; Golub, 1979). Such work derives its motivation from the fact that laryngeal activity often provides a sensitive reflection of abnormality and/or immaturity of the central nervous system.

The recent work of Golub (1979) is particularly promising in this regard. Computerized acoustical analysis was applied to infant cries in order to extract values for formant frequencies, F_0, nasalization, and other parameters believed to be important in distinguishing the cries of normal and ill infants. By comparing actual cry patterns in 2- to 4-day-old infants to a model for normal crying, Golub was able to distinguish normal infants from those known to be ill with jaundice or respiratory distress. The acoustical classification of infants into normal vs. ill was not perfect, but further refinements of this diagnostic test may offer a means of screening for a variety of infant abnormalities during routine hospital testing. In Golub's study, for example, the cries of infants were obtained during a standard heel prick applied for blood testing to detect phenylketonuria (PKU).

Specific characteristics of the infant cries may also be used to diagnose babies who are susceptible to Sudden Infant Death syndrome (SIDS). Golub (1979) has analyzed the crying patterns of infants who later succumbed to SIDS and has found a distinctive pattern of crying that may reflect an anatomical deformity that increases an infant's vulnerability to suffocation (see also Colton and Steinschneider, 1979). In a similar vein, research with infant crying is being extended to examine acoustical characteristics that might reflect certain types of infant brain damage, bacterial meningitis, possibly even congenital deafness. However, it is still too early to determine how well the technique can be applied to detect such disorders, and, in particular, how valuable are the acoustic characteristics of F_0 in such detection.

In addition to crying, F_0 analyses have also been conducted on the nondistress vocalizations of infants. Laufer and Horii (1977) found that the F_0 for a group of four infants averaged about 355 Hz during the first 24 weeks of life, with major variations in F_0 attributed to the infant's activity level and changing articulatory repertoire.

As a young child begins to acquire language, his vocalizations consist primarily of one-word utterances, or holophrases. As the child begins to produce two-word utterances, an F_0 pattern emerges in which the F_0 of the second word typically contains a greater terminal F_0 fall value than that of the first word, analogous to adult speech (Branigan, 1977). Indeed, the terminal effect observed for two-word utterances of young children has been used to provide an operational measure of distinguishing valid two-word utterances from adjacent single word utterances (Stokes and Branigan, 1978).

The F_0 of early child speech also varies as a function of mother–child interaction (e.g., Weaver, 1976; Donahue, 1976; Menn and Boyce, 1978). For example, the F_0 of parents' speech to young children typically contains higher values than speech directed to other adults. It has been suggested that this feature serves to direct the child's attention to speech (Garnica, 1977). In addition, Gleason and Weintraub (1978), among others, have observed that the adult speaker attempts to match his or her F_0 level to the vocalizations of the child. Adults do not appear to adjust their F_0 level to match the level of individual adult listeners, however (Bernstein and Jeje, 1978). One systematic study of mothers' speech to newborns has been conducted with a group of 24 German mothers (Fernald, 1978). In this study, it was found that the F_0 patterns of mothers' speech were greatly simplified and consistent, typified by unidirectional F_0 trajectories in a given utterance.

As the child begins to master a variety of semantic and syntactic representations, such representations begin to influence the child's patterns of F_0, although the details of this emerging influence have yet to be studied. It does seem clear, however, that the child's mastery of adult F_0 patterns is not complete until at least the age of ten. Cruttenden (1974) has vividly demonstrated this point in a study of British children's intonational patterns during the reading of football (soccer) results. For British adults, the results are read with a complex pattern of obligatory and optional intonational rules for falling and rising patterns of F_0 (see also Bonnet, 1980). The score of the home team is always read first, regardless of the winner of the game, and the intonational pattern produced for the home team's score is often sufficient for an adult listener to determine the winner. The intonational patterns differ depending on whether the home team won, lost, or tied. In a study of 28 boys aged 7–10 years old, Cruttenden found that the development of the complex pattern of intonational rules of reading football results was still incomplete by the age of ten.

Between the age of 5 and the onset of puberty, the mean F_0 of the child's speech averages about 270 Hz (e.g., Cornut, Riou-Bourret, and Louis, 1971). Between the ages of 7 and 10, the mean F_0 does decrease slightly, and during this period the average is a bit higher for boys than girls (Hollien, Brown, and Hollien, 1978). As we all know from personal experience, more dramatic changes in F_0 accompany puberty. For members of the Vienna Boys' Choir, the lowering in F_0 that accompanies puberty spells the end of an illustrious singing career and precocious world traveling. For most of us, the changes in F_0 in puberty are not quite so traumatic, but they do represent one of the most visible signs of this maturational stage. A study of adolescent females aged 11, 13, and 15 has charted the course of these F_0 changes at puberty (Duffy, 1970). For males, it appears that most of the decrease in mean F_0 occurs between the ages of 14 and 18 (Curry, 1940).

Systematic changes in F_0 may also accompany adulthood. Sometimes, these changes are attributable to environmental rather than maturational factors. A study of smoking vs. nonsmoking women showed significantly lower mean F_0 for the smokers, in agreement with intuition (Gilbert and Weismer, 1974). Otolaryngeal examination of the smokers in this study revealed some abnormality in the appearance of the larynx for 87% of the smokers. Consideration of the speakers' age also suggested that menopause may be a contributing factor to the lowering of mean F_0 in adult women. In this case, then, both environmental and maturational factors may operate to produce lowered F_0 in later adulthood. Environmentally produced abnormalities in the larynx have also been implicated in a study of the speech of miners with black lung disease (pneumoconiosis) (Gilbert, 1975).

In a study of 175 male speakers ranging in age from 20 to 89 years, Hollien and Shipp (1972) showed that mean F_0 declines by age decade from 20 to 40 but rises from 60 through the 80s. The decrease in mean F_0 until middle age seems to reflect a maturational factor involving the anatomy of the larynx. In particular, the vocal folds increase in thickness during these years. On the other hand, the rise in mean F_0 after age 60 may be attributed to general muscular atrophy that reduces the thickness of the vocal folds and increases their stiffness (Ferreri, 1959).

In summary, the average F_0 values produced by a speaker vary over a considerable range as a function of the speaker's age. The F_0 values are typically highest in infancy and decline in childhood, with the most noticeable drop accompanying puberty. Thereafter, F_0 values continue to decline somewhat from ages 20 to 40 but then rise slightly from 60 to 80. While these changes reflect maturation for the vocal folds and associated musculature, further changes in mean F_0 during adulthood may be produced by environmental factors that act to alter the normal maturational process.

In addition to describing the course of changes in mean F_0 throughout a speaker's lifetime, systematic studies of F_0 at various developmental stages may provide a concrete, real-time tool for investigating a variety of issues about the speaker's general mental condition. This tool is already being applied successfully as a diagnostic for certain infant abnormalities that are otherwise quite difficult to detect, as noted earlier. In addition, measurement of F_0 may provide an effective technique for probing the child's later development of semantic, syntactic, and phonological components of language. The present study has enabled us to identify a number of F_0 characteristics of adult speech that allows one to make inferences about the nature of the speaker's internal linguistic code, and an investigation of how these F_0 characteristics emerge in children's speech should provide one basis for assessing the child's development of such coding.

Pathology

Abnormal F_0 may be produced for a variety of pathologies, including those involving central or peripheral pathology of the language faculty itself as well as more global disorders. Here we shall discuss a wide range of pathologies, beginning with those that involve language selectively.

Aphasia

Aphasia may be produced when an individual suffers a lesion of the language-dominant cerebral hemisphere. Of the various types of aphasia, Broca's and Wernicke's aphasias are most readily distinguishable on the basis of their speech prosody. Broca's aphasics suffer a localized lesion in an anterior portion of the language-dominant cerebral hemisphere, and the speech of such patients is effortful and nonfluent (Goodglass and Kaplan, 1972). In contrast, Wernicke's aphasics suffer a lesion in a posterior portion of this hemisphere, and their speech is generally fluent but semantically empty.

Recently, actual measurements of F_0 and timing have been obtained for both Broca's and Wernicke's patients using the computerized techniques employed in this study. In a study of Broca's speech (Danly, deVilliers, and Cooper, 1979), F_0 and timing measurements were obtained for two-word utterances with both spontaneous speech and in a controlled oral reading experiment. The results showed that, despite interword pauses of up to 5 sec, Broca's patients exhibited two normal sentential properties of F_0, namely, declination of peak F_0 from Word 1 to Word 2 and a greater terminal fall in F_0 from peak to valley on Word 2 vs. Word 1. Thus, while Broca's speech is certainly quite dysprosodic in comparison with normal or Wernicke's speech, it does retain at least rudimentary features of normal sentential F_0 characteristics.

In a study of Wernicke's speech (Cooper, Danly, and Hamby, 1979), a controlled oral reading experiment was conducted using sentences from Study 2.2. The speakers exhibited F_0 declination for these sentences, but the precise form of their declination did not adhere to the Topline Rule. In particular, the Wernicke's patients produced abnormally high values of *P1*. Additionally, the effect observed in Chapter 2 for higher first peaks in long vs. short sentences, also reported by McAllister (1971) and O'Shaughnessy (1976), was not observed for these patients, suggesting that their look-ahead ability is more restricted than normal. The presence of phoneme substitution errors (LITERAL PARAPHASIAS) in their speech did not seriously perturb the form of their F_0 declination, however, indicating that the programming of declination proceeds in a manner that is largely independent of proper phoneme selection.

This latter finding leads to the inference that the programming of declination and proper phoneme selection are probably independent in the speech of normal speakers as well. In this sense, the study of aphasia seems to provide a unique means of assessing the degree to which various components of normal speech information flow are independent of one another (see Cooper and Zurif, 1981, for an update on this research).

Taken in combination, the results for the studies of F_0 in Broca's and Wernicke's patients indicate that the speech of Broca's patients is not completely devoid of rudimentary aspects of normal sentential characteristics of F_0, despite the fact that their speech is marked by numerous interword pauses; on the other hand, the speech of Wernicke's patients does not display the full range of normal F_0 attributes found with normals, despite the fact that their spontaneous speech exhibits many more of the features of normal prosody than that of Broca's aphasics.

Speech Training of Hearing-Impaired Children

For those of us with normal hearing, it is hard to imagine the difficulty of the challenge faced by the hearing-impaired child in attempting to learn language, being deprived of normal auditory input from parents and friends during the most crucial early years of language development. The speech of such children understandably exhibits a variety of abnormalities, including inappropriate prosody. Typically, the mean F_0 for such children is higher than normal, and the range of F_0 values is either too narrow or too broad and erratic (for a review, see Nickerson, 1975). In addition, the speech of deaf children is often marked by abnormally abrupt changes in F_0, degrading intelligibility (e.g., McGarr and Osberger, 1978; Parkhurst and Levitt, 1978). The speech of adult deaf speakers, however, is typically marked by less variation in F_0 compared to the speech of normals (Hood and Dixon, 1969).

Attempts to train improvements in the F_0 of deaf children have relied on tactile or visual feedback methods. Using these methods, it is possible to train some deaf children to produce nearly normal mean F_0 patterns. With tactile feedback, the deaf speaker can monitor his own F_0 by perceiving the location of a mild vibratory stimulus applied to the hand (Stratton, 1974). In the case of visual feedback, the speaker can monitor F_0 by viewing a computerized visual display of his F_0 output (Boothroyd, 1973; Houde, 1973; Watanabe and Okamura, 1976). A game can be played in which the child is reinforced by producing an utterance in which the F_0 pattern falls within a specified range, designated on the display as a roadway, for example. The value of F_0 produced by the child at a given point in time may be displayed as a car, and the child's task is to make the car stay on the roadway throughout the utterance. With a game of this sort, the deaf child can learn to avoid abnormally wide variations in F_0 and also develop attributes such as declination that are commonly found in normal speech.

Profoundly deaf children may also be capable of learning appropriate F_0 patterns via amplified auditory feedback. The audiograms of such children typically show that their most intact hearing occurs for frequencies below 500 Hz, and a signal detectability analysis has shown that profoundly deaf children can discriminate between statements and questions that are low-pass filtered at 510 Hz (Blackwell, Engen, Engen, and Hirsch, 1979). This ability may be attributed to information conveyed by timing and intensity attributes in addition to F_0.

Inappropriate use of F_0 is also found in the speech of children who are hard of hearing but not profoundly deaf. Some brands of hearing aid distort the low frequency information conveyed by F_0 (Sudo, 1973). Since a child wearing such an aid cannot perceive F_0 patterns properly (whether the contours are self-generated or generated by another speaker), he has little opportunity to formulate an adequate mental representation of normal F_0 attributes.

Autism

Childhood autism is characterized by a substantial language deficit and related cognitive problems, although the etiology of this disorder remains obscure (for a review, see Rutter and Schopler, 1978). A common feature of autistic speech in conversation involves echolalia—the inappropriate repetition of another person's speech. Recently, the F_0 patterns of echolalic speech were examined to determine whether prosodic attributes of another person's speech are faithfully mimicked by the autistic child in response to *yes–no* and *wh* questions (Paccia-Cooper and Curcio, 1980). For a group of five children with autistic diagnoses, about half of the echolalic responses to questions exhibited contrastive rather than imitative F_0 patterns. Thus, for example, a child might echo a *yes–no* question that originally ended with a rising F_0 pattern by repeating the words of the question but ending with a falling F_0 pattern. For autistic children with relatively high-level comprehension ability, this type of contrastive F_0 pattern in echolalia, in combination with other evidence from this study, suggested some measure of semantic processing. For echoes in which the rising F_0 pattern of the question was imitated, independent evidence suggested that such echolalia served simply as a primitive turn-taking device or as a rehearsal strategy for the child. The results of the study indicate that the prosodic and lexical aspects of echolalia can be dissociated in the speech of some autistic children.

Laryngeal Pathology

Abnormal F_0 may also result from anatomical and physiological impairment within the larynx itself. For example, variation in F_0 is significantly reduced for speakers with vocal fold paralysis, although the mean F_0 in such speakers is normal (Murray, 1978).

A very different type of language pathology involves a peripheral disorder resulting from surgical removal of the larynx. Laryngectomized patients produce F_0 patterns by holding a mechanical vibrator against the esophagus. Such speakers produce lower F_0 values than normal, but proficient esophageal speakers are capable of producing the patterns of F_0 that are pertinent to intelligibility (Hoops and Noll, 1969; Shipp, 1967; Filter and Hyman, 1975).

Global Pathologies

In addition to pathologies that involve selective impairment of the language faculty, a variety of other disorders may be accompanied by abnormal F_0 in speech production, although the results of testing are somewhat mixed. Mongoloid children with Down's syndrome produce higher mean F_0 than normal children of the same age group in a study of 5- and 6-year-olds (Weinberg and Zlatin, 1970), although no such difference was found in another study (Montague, Brown, and Hollien, 1974). A comparison of F_0 values in mentally retarded and normal adults also failed to show significant differences in the mean F_0 or in F_0 variability (Neeley, Edson, and Carlile, 1968). A study of English adult female schizophrenics and normals showed significantly larger variability in F_0 during oral reading for the schizophrenics, as well as a somewhat higher mean F_0 (Saxman and Burk, 1968). On the other hand, a study of French adolescent female schizophrenics showed a reduced F_0 range for patients compared to normals (Chevrie-Muller, Dodart, Seguier-Dermier, and Salmon, 1971).

While the relationship between mean F_0 and schizophrenia remains obscure, it does appear that mean F_0 distinguishes psychiatric patients exhibiting abnormally low affect (so-called "flat" affect) from patients having normal affect. In one study (Rice, Abroms, and Saxman, 1969), F_0 was measured for patients while reading a rehearsed, nonemotional passage and while telling an emotion-evoking story. It was found that patients who were predetermined to exhibit flat affect produced higher mean F_0 while reading the nonemotional passage, whereas patients having appropriate affect produced higher mean F_0 while telling the emotional one. In addition, the patients having flat affect showed a decrease in heart rate and body movement during the affective storytelling compared with their performance while reading the nonemotional passage. This latter result suggests that patients with flat affect actively inhibit their arousal system when confronted with affect-evoking stimulation.

Personality and Emotion

A speaker's personality and particular emotional state also exert an influence on F_0. Accordingly, listener's perception of personality and emotional traits are often influenced by voice pitch. It is well known that a speaker's "tone of voice" often provides a listener with as much information about the speaker's emotional state as does the semantic content of his utterances.

In recent years, studies have pointed to the influence of a variety of personality traits on F_0 in speech production, including, for example, introversion vs. extroversion (Trimboli, 1973), and depression vs. schizophrenia (Swanson, 1977). In addition, F_0 appears to be influenced by personality traits revealed during interpersonal interaction, including traits like confidence (Scherer, London, and Wolf, 1973) and deception (Eckman, Friesen, and Scherer, 1976).

The short-term emotional state of the speaker also influences F_0 patterns. For example, heightened emotional stress typically produces an increase in mean F_0, attributable to an increase in vocal fold tension (Hecker, Stevens, von Bismarck, and Williams, 1968; Levin and Lord, 1975; Williams and Stevens, 1972; Hauser, 1976; Utsuki and Okamura, 1976). This relationship was pointed out most dramatically in a study of the speech of pilots just prior to anticipated aircraft accidents (Utsuki and Okamura, 1976). Of the four pilots studied, all showed increases in mean F_0 as the moment of the crash approached, but the more experienced pilots disproportionately increased the standard deviation of F_0 in order to maintain voice intelligibility. In a study of another harrowing event, Streeter, Krauss, Apple, and MacDonald (1978) examined the F_0 values for telephone conversations of a systems operator at the Consolidated Edison electrical company, beginning one hour before the 1977 New York blackout. During this emotionally stressful hour, the operator's mean F_0 systematically decreased. Yet, when listeners were asked to judge the operator's stress while listening to the speech, they showed a positive correlation between mean F_0 and perceived stress (see next section on Perception). The systematic long-term decline in the speaker's mean F_0 during this exasperating hour stands in contrast to the sharp increase in mean F_0 for the Japanese airline pilots facing the prospect of an imminent plane crash. Certainly, the pilots faced a more highly stressful situation within a much shorter time span than the electrical operator, and this difference in stress and time span may account for the disparate results obtained in these two studies.

Perception

The perception of F_0 may provide information about a variety of issues, including the utilization of F_0 by the listener in decoding linguistic and nonlinguistic aspects of speech. In addition, perception of F_0 may be studied from the standpoint of the lateralization of perceptual processing in the brain.

Attributes of F_0 may be utilized by the listener in decoding a variety of types of linguistic information, including the identity of consonants as well as the identity of major clause and phrase boundaries. At the segmental level, differences in F_0 may serve to distinguish voiced from voiceless consonants. Among the stop consonants, listeners are more likely to perceive a word-initial consonant as voiced when the F_0 contour during the first 50 msec of the following vowel contains a rising as opposed to falling pattern (Fujimura, 1971; Haggard, Ambler, and Callow, 1970). Similar results have also been obtained for voicing distinctions between the fricatives [z] and [s] (Massaro and Cohen, 1977). The perceptual cues to voicing for the consonants mirrors the F_0 pattern found in speech production for these consonants (see Chapters 1 and 4).

In a well-known study, Fry (1955) showed that listeners perceive higher F_0 as stressed in selecting between words differing only in stress pattern, e.g., *con*vict and con*vict*. Surprisingly, no systematic studies have been conducted to determine whether the perception of stress or F_0 is influenced by factors such as inherent vowel quality or the structure of neighboring segments, both of which are known to influence F_0 in speech production.

Variations in F_0 pattern in word and sentence contexts may be utilized by the listener to determine whether the utterance is a question or a statement. In a study of 22 Swedish and 16 American listeners, falling terminal F_0 patterns on a word yielded judgments that the utterance was a statement, while rising F_0 patterns on the same word yielded judgments that the utterance was a question (Studdert-Kennedy and Hadding-Koch, 1973). In a related study, Hadding-Koch and Studdert-Kennedy (1974) found that listeners perceived a level terminal F_0 as being produced by a speaker talking to himself when such a terminal pattern was preceded by a level, low to moderate value of F_0 during the utterance.

In another study of synthetic speech, Majewski and Blasdell (1969) found that the value of F_0 terminating the utterance was another factor in listeners' judgments of statements vs. questions. This factor was significantly related to such judgments in both English and Polish listeners, but more so for the Polish listeners. A high terminal F_0 produced judgments of questions, whereas a low terminal value yielded judgments of statements, independent of the F_0 trajectory near the end of the utterance.

A terminal fall in F_0 may be utilized by listeners in detecting the presence of a major syntactic boundary within an utterance (de Rooij, 1979; Streeter, 1978). Streeter (1978), for example, has shown that variations in F_0 can also aid listeners in decoding ambiguous algebraic expressions of the form "(*A* plus *E*) times *O*" vs. "*A* plus (*E* times *O*)." Although the acoustical properties of these F_0 variations were not reported in this study, it is likely that fall–rise patterns of F_0 (Chapter 3) led to the perception of a major phrase boundary in such expressions.

For the most part, studies have yet to be conducted to assess the role of such F_0 patterns during the perception of fluent sentential speech under normal circumstances (for extended discussion, see Cooper and Paccia-Cooper, 1980). We might well expect that the contribution of F_0 to the listener would be most apparent when the speech signal itself was degraded by noise or other competing stimuli. In such circumstances, it is altogether likely that the speaker accentuates certain F_0 attributes to aid the listener in decoding (see also de Rooij, 1979).

Listeners' judgments of the "naturalness" of F_0 contours have been used to investigate the appropriateness of various F_0 features for speech synthesis. Delgutte (1978), for example, examined syntactic and focus features of F_0 in French and found that the features judged to be most natural were the features that also appeared most reliably in spoken French. These features were thus deemed to be among the most important for implementation in a speech synthesis program for French.

Aside from their function in decoding linguistic information, F_0 patterns also aid the listener in decoding nonlinguistic aspects of an utterance. For example, continuity of F_0 seems to provide the listener with information that the speech is being produced by a single speaker (Darwin and Bethell-Fox, 1977). Large differences in mean F_0, on the other hand, yield the percept of different speakers (Ades, 1977; Nooteboom, Brokx, and de Rooij, 1978). Mean F_0 serves as an indicator of whether the speaker is male or female, with higher mean F_0 values being perceived as associated with female speakers, other factors being equal (e.g., Coleman, 1976). The range of F_0 also provides information about perceived emotional state of the speaker, with higher ranges correlated with perceived emotional intensity (Huttar, 1968).

In addition, F_0 provides the listener with clues about the speaker's personality. In an experiment in which the F_0 of male speakers' voices were raised or lowered by 20% electronically, Apple, Streeter, and Krauss (1979) found that listeners perceived the speech having higher F_0 values as being uttered by a speaker who was less truthful, less persuasive, more nervous, and weaker than normal. Similar effects were obtained when the speech was electrically slowed, except that the speakers were perceived as being less nervous than normal.

In the study of Apple et al., the manipulation of raising the F_0 values was accompanied by an increase in the variance of the F_0 values. However, as the authors themselves point out, it is unlikely that their results can be attributed to the influence on F_0 variance. In another study, Brown, Strong, and Rencher (1974) manipulated mean F_0 and F_0 variance independently and found that an increase in the mean F_0 or a DECREASE in F_0 variance yielded a reduction in the perceived "competence" and "benevolence" of the speakers. According to these results, an increase in F_0 variance tends to heighten perceived competence, unlike the effects obtained by Apple et al. The earlier results thus seem to be attributed primarily to the increase in mean F_0.

F_0 also influences a listener's judgments about the sexual characteristics of a speaker. Among males, speakers with high mean F_0 were judged more effeminate than speakers with lower values (Terrango, 1966), even though a study of mean F_0 values in homosexual vs. heterosexual males showed no difference in this value (Lerman, 1969). Among females, speakers with low mean F_0 were judged more virile than speakers with high mean F_0 (e.g., Damste, 1967).

The perception of F_0 also provides clues about a speaker's age. Horii and Ryan (1975) found that the perceived age of a group of 57 adult male speakers was positively correlated with their mean F_0, even though, as pointed out by Hollien and Shipp (1972), actual age is positively correlated with mean F_0 only in the older age range from 60 through the 80s (see earlier section on Language Development).

In one series of experiments, the perception of an F_0 attribute itself has been the object of study. In these experiments, listeners were asked to judge the perceived pitch of a syllable as a function of its position in a short utterance (Breckenridge and Liberman, 1977; Pierrehumbert, 1979b). The results showed that listeners normalize their pitch perception to compensate for F_0 declination even when such declination was not present in the acoustic stimulus. For example, listeners judged a syllable appearing near the end of the utterance to have a higher pitch value than a syllable appearing earlier which contained an identical F_0 value. In further work, it would be interesting to determine whether the listener's normalization for declination mirrors the Topline Rule presented in Chapter 2 of this study or whether such normalization is represented more grossly. In addition, one could determine whether individual differences in the perceptual normalization for F_0 declination are highly correlated with individual differences in speech production of this attribute. This latter question could be pursued along the same lines as a study of individual differences in speech timing (Klatt and Cooper, 1975).

In earlier research, perceptual normalization was revealed for acoustic variables such as formant frequencies (Ladefoged and Broadbent, 1957) and speech rate (Summerfield and Haggard, 1972). Normalization for these factors is critical for the decoding of segmental information about the phonemic identity of vowels and consonants in some circumstances. The discovery of normalization for F_0 declination is somewhat more surprising, since in English the perception of pitch levels does not play a primary role in decoding phonetic segments.

Differences in the pattern of F_0 may also give rise to differences in the perceived duration of a speech sound. Lehiste (1976), for example, has shown that a synthetic vowel is perceived as being longer when a rise–fall pattern of F_0 is superimposed on the vowel than when the F_0 pattern is level.

A different question about perception concerns the general location in the brain where perceptual processing of F_0 is accomplished. In normal adults, the general task used most often to infer the location of cerebral processing involves dichotic listening, in which two stimuli are presented to the listener simultaneously, one to each ear. If listeners exhibit a preference for correctly identifying stimuli presented to one ear, it is inferred that the critical stimulus attribute is processed by the contralateral cerebral hemisphere. Thus, for example, consonants are typically identified better when presented to the right ear of a right-handed listener, suggesting that the processing of consonants takes place primarily in the left cerebral hemisphere (e.g., Studdert-Kennedy and Shankweiler, 1970). In the case of F_0, the location of processing appears to depend critically on the perceptual task. Haggard and Parkinson (1971) have shown that variations in F_0 give rise to a right-ear advantage in dichotic listening when this attribute is used to distinguish different stop consonants, whereas variations in F_0 yield a left-ear advantage when used to distinguish the emotional tone of a sentence. It thus appears that F_0 information may be processed by either cerebral hemisphere, the choice depending on the critical function of this information in the task at hand.

Memory

Relatively little is known about human memory for F_0 information, despite the fact that long-term memory for F_0 may play some role in our ability to recognize familiar speakers by voice alone over the telephone. Studies have shown that short-term memory for F_0 information is relatively stable over a 2–8 sec period (e.g., Cole, 1973; Springer, 1973; Wolf, 1977). It would be interesting to study long-term F_0 memory, as part of an overall program to test subjects' ability to identify speakers on the basis of limited prior exposure. Just as we possess a highly accurate long-term memory representation of certain odors (e.g., Engen and Ross, 1973), faces (Marslen-Wilson and Teuber, 1973), and entire events (Brown and Kulik, 1976), so we might have a very good representation for F_0 information insofar as such information is useful for speaker identification, particularly for speakers whose nonacoustic characteristics are also prominently represented in memory. For example, internal representation of the F_0 patterns of John F. Kennedy's voice might be comparable in accuracy to the representations of his face or hand gestures. The ability to form mental images of F_0 and other acoustical properties of a speaker's voice also seems highly developed, and systematic study of this ability would bear relevance to a variety of issues currently addressed primarily from the standpoint of visual imagery (e.g. Kosslyn, 1980; Hinton, 1981).

Speech Synthesis-by-Rule

In a program to synthesize speech by rule, a phonetic transcription of an utterance is converted into a speech output. Synthesis-by-rule finds widespread application in man–machine communications and forms a component of a reading machine for the blind (e.g., Allen, 1973; Caldwell, 1978). At present, a variety of synthesis programs have been devised for English and other languages, including French, Dutch, Swedish, and Japanese, among others. These programs are aimed chiefly at practical applications, but they also serve as a concrete proving ground for theoretical formulations in human speech production. As observed in Chapter 2, our own thinking on F_0 declination has been influenced by the synthesis-by-rule research conducted by Pierrehumbert (1979a), for example.

Current synthesis programs typically suffer from improper rhythm and fundamental frequency contours. This drawback is not surprising in light of the fact that our knowledge about these global factors is just beginning to approach the same degree of sophistication as our knowledge about the production of individual phonetic segments. However, improvements in the rhythm and F_0 of synthetic speech can be implemented on the basis of what is now being learned about how speakers produce these attributes.

In the case of F_0, the characteristics of declination and fall–rise patterns seem particularly relevant to the generation of natural-sounding synthetic speech. Synthesis programs for English by Mattingly (1968), Flanagan, et al. (1970), Olive (1975), O'Shaughnessy (1976), Maeda (1976), Klatt (1979), and Pierrehumbert (1979a), among others, have incorporated some aspects of these characteristics already. These features have also been included in synthesis programs for French (Choppy, 1978; Delgutte, 1978), Swedish (e.g., Carlson, Granstrom, Lindblom, and Rapp, 1972; Carlson and Granstrom, 1973; Gårding, 1977), and Dutch (Cohen and t'Hart, 1967; t'Hart and Cohen, 1973). The first systematic study of F_0 declination was actually undertaken during a synthesis-by-rule project by the Dutch investigators Cohen and t'Hart. The synthesis of Swedish is particularly interesting because declarative sentences are typically produced with a terminal rise in F_0, yet F_0 declination seems to be observed throughout the earlier portions of the utterance (Gårding, 1977).

Most of the early synthesis-by-rule programs incorporated the features of a male speaker. Recently, however, attempts have been made to synthesize speech having female characteristics as well (e.g., Goldstein, 1972; Yasuhiro and Ozeki, 1976), and a major issue involves whether a single scaling factor for F_0 can be used to derive female F_0 patterns from male patterns and vice versa. In general, male F_0 values have been multiplied by a constant factor to mimic female values. Goldstein (1972), for example, employed a scaling factor of 1.7, based on the steady-state vowels for female and male speech as determined by Peterson and Barney (1952). For the vowels in their corpus, the scaling factors ranged from 1.64 to 1.73 with a mean of 1.69. Whereas the Peterson and Barney data included only one measured value of F_0 for each utterance, the data obtained in our experiments on fall–rise patterns show that the use of a single scaling factor is appropriate for each of two or more disparate measured points in the region of clause and phrase boundaries. Separate analyses of male and female F_0 values for Study 3.1 on fall–rise patterns indicate that a scaling factor of about 1.8 serves as a very good approximation for synthesizing female F_0 based on male values (see Cooper and Sorensen, 1977, Table IV). The results of this analysis illustrate a general principle observed throughout the data of our experiments; namely, that male and female speakers produce the same general F_0 patterns, the major difference being that the absolute female values of F_0 are generally 1.8 times higher than male values. The similarity in the patterning of F_0 for the two sexes greatly simplifies the problem of converting from one sex of speaker to another in speech synthesis. The high degree of similarity in the patterns of F_0 attributes for the two sexes is quite understandable in light of the non-sex-specific factors that are believed to exert primary influence over these attributes, including both linguistic and extralinguistic variables.

While the implementation of a simple scaling factor certainly plays a large role in producing synthetic speech that mimics male vs. female voices, this is by no means the only factor required for a convincing synthesis of female speech. As noted by Monsen and Engebretson (1977), differences in the shape of individual glottal cycles also accompany male vs. female speech, and a high degree of naturalness for synthesis of a female voice can only be achieved by taking into account the particular features of the glottal source for females. For example, the opening and closing phases of the glottal cycle occupy about the same amount of time for female speech. For male speech, on the other hand, the closing phrase occupies only about 20–40% of each glottal period. In addition, the slope of the glottal spectrum is steeper for the female voice at any octave. The more general spectral envelope, however, is about the same for male and female speech, since the steeper glottal spectrum for females is counterbalanced by the higher F_0 and lower intensity of their speech. In general, the differences between male and female glottal spectra can be accounted for in terms of anatomical differences in the length of the vocal folds themselves, with shorter folds for females acting more like single masses and the longer folds for males acting like two-mass bodies, as reviewed in Chapter 1 (Ishizaka and Flanagan, 1972; Monsen and Engebretson, 1977).

Speech Recognition by Machine

In addition to speech synthesis, communications engineers have attempted to develop computer systems that can recognize human speech as part of an overall effort toward man–machine communication by natural language (for reviews, see Reddy, 1974; Klatt, 1978). In at least one particular program of speech recognition, fall–rise patterns of F_0 have played a central role (Lea, 1973, 1979).

In Lea's program, fall–rise patterns are detected at an initial stage and form the basis for the machine's location of major syntactic boundaries. As expected from Lea's own research with human speech as well as from the results of Chapter 3, the machine correctly identifies a very large proportion of major syntactic boundaries on the basis of information provided by fall–rise patterns. Once these boundaries are located, a routine for phoneme recognition is directed toward segments lying in the immediate region of the boundaries. The advantage of initially directing the phoneme recognition routine to this region lies in the fact that cross-word conditioning effects which normally hamper correct recognition seldom apply across such boundaries, as observed at the outset of Chapter 4. The phoneme recognition device is thus directed initially to the location in the speech wave where phonemic information is more reliably represented in the acoustic signal, minimizing the dire cumulative effects that initial errors in phoneme recognition produce at subsequent stages of the recognition system. While human listeners normally recognize syllables in a left-to-right fashion as the speech is produced, the machine's task is aided by the strategy of initial recognition of phonemes in the region of major syntactic boundaries.

Speaker Recognition

In the past, a number of efforts have been aimed at refining the technique of VOICEPRINT IDENTIFICATION. This procedure, supposedly analogous to fingerprint identification (Kersta, 1962), held promise as a source of evidence in a variety of legal circumstances. According to Kersta, a speaker's identity can be established from a spectographic examination of the voice. Generally speaking, the use of such an acoustical analysis to identify a speaker is quite difficult, and the validity of such evidence in legal circumstances has been strongly questioned (Bolt, Cooper, David, Denes, Pickett, and Stevens, 1970, 1973).

Nevertheless, there exist situations in which speaker recognition is useful and reliable. Such situations rarely, if ever, make use of visual matching of spectograms. Rather, the extraction and comparison of the appropriate speech properties is performed by computer. One such area of speaker recognition is termed SPEAKER VERIFICATION. In this task, the goal is to determine if a given person is who he claims to be, based on comparison of a current sample of speech with previously recorded speech from the same individual. The mean and distribution of a speaker's F_0 values have been found to be useful in such verification schemes (Doherty, 1976; Lummis, 1972; Wolf, 1972; Hollien, Johnson, and Doherty, 1978; Ichikawa, Nakajima and Nakata, 1978). One particularly useful application of speaker verification involves entry control to restricted facilities (Hydrick and Doddington, 1978). The type of F_0 patterns examined in Chapters 2 and 3 may eventually show evidence of being speaker-dependent to some degree, and if so, could be helpful in both verification and recognition.

Animal Communication

Patterns of F_0 play an important role in the communicative systems of many nonhuman species. In birdsong, studies have revealed that melodic patterns convey information about mating, food, and other significant factors. Acoustical analyses of birdsong have indicated that F_0 inflections contribute to this type of communication in at least some types of birds. A system of communication based largely upon F_0 patterns appears to exist for the species *Agapornus roseicolis,* for example, as indicated by a study in which natural and synthetic vocalizations were correlated with birds' behavior (Sedlacek, 1971). Large-scale acoustical analyses of birdsong still remain to be conducted for the most part, but even a casual examination of spectographic traces of song patterns suggests that F_0 movements play a role in both the semantic and syntactic attributes of this form of communication.

The extent to which birds utilize F_0 in the same ways as humans remains to be determined. In preparing this section, we wrote to a number of leading researchers of birdsong for their opinions on this issue.[1] Generally, the respondents indicated that, on the basis of informal listening, birds often exhibit F_0 declination, but in many cases the very last note of a phrase is equal to or higher in F_0 than the preceding note, depending on the species.

While most species of birds produce species-specific song patterns, some species (e.g., mynah birds) can also imitate human speech. The F_0 patterns of imitations have been studied in detail for one trained mynah bird (Klatt and Stefanski, 1974), and the results show a high degree of consistency between the F_0 patterns of bird and human. The general F_0 patterns of human speech are better preserved than are the absolute F_0 values by the mynah bird. The bird seems capable of imitating the F_0 patterns produced by the human larynx by means of a pair of external labia in the bird's syrinx.

In addition to birdsong, F_0 has been studied to some extent in the vocalizations of dolphins (e.g., Lilly, 1963, 1965), toads (e.g., Schneider, 1976), rats (e.g., Ghiselli and LaRivière, 1977), and bees (e.g., Es'kov, 1969), among other species. For the most part, these studies have characterized the general range of F_0 patterns produced by the species and have indicated how variations in F_0 convey different meanings. Phrasal units can sometimes be independently determined in the vocalizations of higher mammals such as dolphins and whales, and it would be interesting to study the influence of phrasal coding on F_0 patterns in these animals to determine the extent to which the influence bears resemblance to that found in human speech.

[1]We thank Joan Hall-Craggs, R. E. Lemon, Peter Marler, and Keith Nelson for their correspondence on parallels between speech and birdsong.

Other Languages

F_0 studies have been conducted in a variety of other languages, although the majority of such studies do not bear directly on the major questions addressed throughout this study. In Table 6.1, we cite a large (though by no means exhaustive) list of related articles on various languages other than English. In constructing this table, we limited ourselves primarily to cases in which F_0 was either measured acoustically or was used as an independent variable in related experimentation on perception. In a few cases, we cited studies that did not include direct measurements of F_0 but which provide a basis for research in languages for which little or no work on F_0 has been conducted. Readers interested in studies on tone (see later) should consult the extensive table provided by Fromkin (1978, pp. 285–288).

Dialects

In addition to research on other languages, F_0 studies of any given language may be subclassified according to dialect. In English, the differences in F_0 patterns accompanying British vs. American dialects are often quite pronounced (e.g., Crystal, 1969). As noted in Chapter 1, the experimental studies presented in this book were conducted exclusively with native speakers of American English, and there is little doubt that some of our results (e.g., the precise form of the Topline Rule) do not generalize to British dialect. On the other hand, it is expected that many of the more general principles emanating from these studies (e.g., the influence of syntax on F_0 patterns) will appear in all dialects. In addition to the major distinction between British and American English, there remains the need for research to determine differences in F_0 patterns for various subdialects of American and British English, typically defined in terms of race or geographical region. Intuition suggests that subdialects exhibit some distinctive characteristics of F_0, but the acoustical studies required to make such a determination have for the most part yet to be conducted.

Tone Languages

Generally speaking, F_0 variations in English are distinctive linguistically for signaling differences in syntax or stress. In tone languages, however, F_0 also plays a role in signaling differences among lexical items and grammatical relations (Lehiste, 1970; Fromkin, 1978). F_0 distinguishes lexical items in a variety of Asian and African languages, including the most commonly spoken tone language, Chinese. Grammatical relations are distinguished by F_0 less frequently than are lexical items, but cases of the former can be found in Serbo-Croatian, for example, where locative and dative relations are distinguished by rising vs. falling tones on the same syllable, respectively.

Table 6.1. F_0 Research on other languages

Cantonese
  Benson, Smith, and Arreaga (1972)

Chinese
  Abramson (1974, 1975)
  Chang (1958)
  Cheng (1968)

Czech
  Janota and Liljencrants (1969)

Danish
  Basbøll (1978)
  Jeel (1975)
  Petersen (1978)
  Thorsen (1978a, 1978b)

Duala
  Lhote (1971)

Dutch
  Cohen and t'Hart (1967)
  Collier (1975)
  Collier and t'Hart (1971, 1972)
  De Rooij (1979)
  Hollien and Hollien (1971)
  Klein, Plomp and Pols (1970)
  t'Hart and Cohen (1973)
  t'Hart and Collier (1978)
  van Geel (1979)
  van Katwijk (1970)
  Willems (1978)

Estonian
  Lehiste (1971)

Ewe
  Clements (1977)

French
  Baligand and James (1973)
  Beeckmans (1973–1974)
  Boë (1973–1974)
  Boë and Rakotofiringa (1975)
  Chevrie-Muller, Dodart, Seguier-Dermier, and Salmon (1971)
  Choppy (1978)
  Contini and Boe (1973)
  Cornut, Riou-Bourret, and Louis (1971)
  Delattre (1963, 1966)

French, continued
  Delgutte (1978)
  Larreau and Boë (1973)
  Levelt, Zwanenburg, and Ouweneel (1970)
  Rigault (1961)
  Serniclaes and Wajskop (1971)
  Vaissière (1974, 1975, 1977)
  Zwanenburg (1970)

German
  Bleakley (1973)
  Delattre (1963)
  Delattre, Poenack, and Olsen (1965)
  Isačenko and Schädlich (1970)
  Scherer (1974)

Gonja
  Painter (1971)

Gwa
  Painter (1974)

Hausa
  Meyers (1974, 1976)
  Wangler (1963)

Hindi
  Kagaya and Hirose (1975)

Hungarian
  Fonagy (1971)
  Magdics (1964, 1969)

Islandsk
  Pétursson (1978)

Japanese
  Akiyama and Yumoto (1966)
  Curry, Snidecor, and Isshiki (1973)
  Fant, Ishizaka, Lindqvist, and Sundberg (1972)
  Fujisaki, Hirose, and Sugito (1976)
  Fujisaki and Nagashima (1967)
  Fujisaki and Sudo (1970, 1971)
  Fujisaki and Sugito (1976)
  Hiki (1966)
  Hiki, Sugawara, and Oizumi (1966)
  Hirose and Ushijima (1978)

Table 6.1., continued

Japanese, continued
- Shimaoka (1966)
- Tatsumi, Kunisaki, and Fujisaki (1976)
- Watanabe and Okamura (1976)
- Yasuhiro and Ozeki (1976)

Lithuanian
- Robinson (1968)

Mandarin
- Howie (1974)

Norwegian
- Efremova, Fintoft, and Ormestad (1965)
- Fintoft (1970)
- Zetterlund, Norstrand, and Engstrand (1978)

Polish
- Hollien and Hollien (1971)
- Jassem (1975)
- Jassem, Morton, and Steffen-Batog (1968)
- Kuzniarz (1967)
- Majewski, Hollien, and Zalewski (1972)
- Mikos (1976)

Russian
- Chistovich (1969)
- Haavel (1975)
- Krovnova (1975)
- Slepokurova (1971)
- Svetozarova (1975)

Scandinavian Tonal Accents
- Öhman (1967)

Serbo-Croatian
- Ivić and Lehiste (1963, 1965, 1967, 1969)
- Lehiste (1961)
- Lehiste and Ivic (1963, 1972)
- Purcell (1976, 1978)
- Rehder (1968)

Spanish
- Delattre (1963)
- Delattre, Olsen, and Poenack (1962)
- Hollien and Hollien (1971)

Spanish, continued
- Kvavik (1974)
- Navarro (1957)
- Olsen (1972)

Swedish
- Alstermark and Erikson (1971)
- Bruce (1975)
- Carlson, Erikson, Granstrom, Lindblom, and Rapp (1974)
- Carlson and Granstrom (1973)
- Carlson, Granstrom, Lindblom, and Rapp (1972)
- Erikson (1973a, 1973b)
- Erikson and Alstermark (1972)
- Gårding (1973, 1977a)
- Gårding, Fujimura, and Hirose (1970)
- Gårding and Lindblad (1973)
- Hadding-Koch (1961)
- Hadding-Koch and Studdert-Kennedy (1963, 1964a, 1964b)
- Hollien and Hollien (1971)
- Löfqvist (1973, 1975)
- Öhman (1967)
- Westin, Buddenhagen, and Obrecht (1966)
- Zetterlund, Nordstrand, and Engstrand (1978)

Tagalog
- Gonzalez (1970)

Thai
- Abramson (1962, 1974, 1975, 1976)
- Gandour (1974a, 1974b)
- Lhote (1971)
- Zee (1978)

Turkish
- Nash (1973)

Twi
- Painter (1976)

Vietnamese
- Lhote (1971)

West African
- Ladefoged (1964)

Yoruba
- Lavelle (1974)

In attempting to characterize the distinctive uses of F_0 in tone languages, linguists have proposed a set of tonal features, including static attributes such as High, Mid, and Low as well as dynamic attributes such as Rising and Falling (e.g., Wang, 1967; Abramson, 1976). For present purposes, however, the most interesting aspect of tonal systems is their capacity for interplay with features of F_0 that are programmed over a domain larger than a single word. An example of such interplay may be found in the phenomenon of tone sandhi, referring to the phenomenon in which the tone of a particular syllable is influenced by the tone of adjacent syllables. At word boundaries, sandhi may be blocked if the word boundary coincides with a major syntactic boundary (e.g., Wang, 1967; Meyers, 1976, p. 117; Clements, 1977), in much the same way that blocking constrains other cross-word conditioning effects, as reviewed at the beginning of Chapter 4. Aside from this observation, however, there is little evidence currently available on tone languages that might bear on the generality of the findings obtained here for American English. This situation is likely to improve with the more widespread application of acoustical analysis to tone language speech. As more information is gathered about the F_0 characteristics in the speech of diverse languages, it will become possible to determine which aspects of F_0 are universal and which are specific to a particular language or language family.

Conclusion

We have seen how acoustical measurements of F_0 provide a means for studying a host of topics in both basic and applied research. The intricacies of F_0 patterns concern not only the communications engineer, the speech pathologist, and the psycholinguist but also the social psychologist and the pediatrician. The realization that the study of F_0 impinges quite directly on such a wide range of other topics confirms the belief that an in-depth study of any single topic inevitably leads one to consider many others. It is fair to say that we have just begun to tap the reservoir of information contained in F_0 patterns from the standpoint of what such patterns can tell us about the speaker's mental operations. And, as more is learned about this topic, opportunities will arise for guiding research on other issues mentioned in this chapter. As indicated here, many other studies on F_0 have considered only gross characteristics of this attribute, such as its mean value and variance over long stretches of speech. Using the results of previous chapters as a guide, it should be possible in future work to obtain more information from F_0 patterns by considering the characteristics of declination, fall–rise contours, and F_0 conditioning effects. But the direction of guidance is expected by no means to be exclusively one-way—advances in applied research may well provide new hints about aspects of normal speech production that would otherwise go unnoticed. Some evidence of this direction has already emerged from studies on such topics as aphasia and speech synthesis-by-rule.

References

Abramson, A. S. The vowels and tones of standard Thai: Acoustical measurements and experiments. *Publication 20 of the Indiana University Research Center in Anthropology, Folklore and Linguistics,* 1962, *9,* 146. (Also in International Journal of American Linguistics, *28,* 2, part II.)

Abramson, A. S. The tones of central Thai: Some perceptual experiments. *Haskins Laboratory Status Report on Speech Research,* 1974, *39–40,* 91–104.

Abramson, A. S. Pitch in the perception of voicing states in Thai: Diachronic implications. *Haskins Laboratory Status Report on Speech Research,* 1975, *41,* 165–174.

Abramson, A. S. Static and dynamic acoustic cues in distinctive tones. *Haskins Laboratory Status Report on Speech Research,* 1976, *47,* 121–127.

Ades, A. E. Source assignment and feature extraction in speech. *Journal of Experimental Psychology: Human Perception and Performance,* 1977, *3,* 673–685.

Akiyama, K., and Yumoto, K. A study of voice identification using Japanese speech. *Study of Sounds,* 1966, *12,* 209–223.

Allen, G. D. Segmental timing control in speech production. *Journal of Phonetics,* 1973, *1,* 219–237.

Allen, J., and O'Shaughnessy, D. Fundamental frequency contours of auxiliary phrases in English. Unpublished Paper. Research Laboratory of Electronics and Department of Electrical Engineering, M.I.T., Cambridge, Mass., 1974.

Alstermark, M., and Erikson, Y. Swedish word accent as a function of word length. *Stockholm Transmission Laboratory Quarterly Progress Status Report,* 1971, *1,* 1–19.

Apple, W., Streeter, L. A., and Krauss, R. M. Pitch and speech rate. *Journal of Personality and Social Psychology,* March, 1979.

Atkinson, J. E. Aspects of intonation in speech: Implications from an experimental study of fundamental frequency. Unpublished Ph.D. Thesis, University of Connecticut, Storrs, Conn., 1973.

Atkinson, J. E. Inter- and intraspeaker variability in fundamental voice frequency. *Journal of the Acoustical Society of America,* 1976, *60,* 440–445.

Atkinson, J. E. Correlation analysis of the physiological factors controlling fundamental frequency. *Journal of the Acoustical Society of America,* 1978, *63,* 211–222.

Baer, T. Investigation of phonation using excised larynxes. Unpublished Ph.D. Dissertation, M.I.T., Cambridge, Mass., 1975.

Baligand, R., and James, E. The intonation of *wh-* questions in Franco-Ontarian. *The Canadian Journal of Linguistics/La Revue Canadienne de L'inguistique,* 1973, *18,* 89–101.

Basbøll, H. A note on boundaries and stress rules in Danish phonology. In E. Gårding,

G. Bruce, and R. Bannert (Eds.), *Nordic Prosody: Papers from a Symposium.* Department of Linguistics, Lund University, 1978. Pp. 65–72.

Beeckmans, R. Influence de la frequence fondamentale sur l'espace perceptif des voyelles. *Université Libre de Bruxelles Rapport d'activities de l'institut de Phonetique,* 1973–1974, *8,* 1–2, 109–128.

Benson, P., Smith, T., and Arreaga, L. Dichotic listening of lexical tone by speakers of Cantonese and English. Paper presented at the 84th meeting of the Acoustical Society of America, Miami Beach, Fla., 1972.

Bernstein, N., and Jeje, S. Higher pitch in baby talk . . . are these trix just for kids? Unpublished Paper, Boston University, Boston, Mass., 1978.

Bierwisch, M. Regeln für die Intonation deutscher Sätze. *Studia Grammatica,* 1966, *7,* 99–201.

Blackwell, P. M., Engen, E. A., Engen, T., and Hirsch, S. C. The perception of intonation in hearing impaired children. Unpublished Paper, The Rhode Island School for the Deaf and Brown University, 1979.

Bleakley, D. The effect of fundamental frequency variations on the perception of stress in German. *Phonetica,* 1973, *28,* 42–59.

Boë, L-J. Etude de l'interaction des caracteristiques intrinseques des consonnes du francais. *Bulletin de l'institut de Phonetique de Grenoble,* 1973, *2,* 1–24.

Boë, L-J. Quelques remarques et precisions concernant l'etude statistique de la frequence laryngienne. *Bulletin de l'institut de Phonetique de Grenoble,* 1975, *4,* 67–84.

Boë, L-J., and Rakotofiringa, H. A statistical analysis of laryngeal frequency: Its relationship to intensity level and duration. *Language and Speech,* 1975, *18,* 1–13.

Bolinger, D. Intonation as a universal. In *Proceedings at Linguistics IX.* The Hague: Mouton, 1964. Pp. 833–844.

Bolinger, D. Accent is predictable (if you're a mind-reader). *Language,* 1972, *48,* 633–644.

Bolt, R. H., Cooper, F. S., David, E. E., Jr., Denes, P. B., Pickett, J. M., and Stevens, K. N. Speaker identification by speech spectograms: A scientist's view of its reliability for legal purposes. *Journal of the Acoustical Society of America,* 1970, *47,* 597–612.

Bolt, R. H., Cooper, F. S., David, E. E., Denes, P. B., Pickett, J. M., and Stevens, K. N. Speaker identification by speech spectograms: Some further observations. *Journal of the Acoustical Society of America,* 1973, *54,* 531–534.

Bonnet, G. A study of intonation in the soccer results Wolverhampton Wanderers 2 Nottingham Forest ? *Journal of Phonetics,* 1980, *8,* 21–38.

Boothroyd, A. Some experiments on the control of voice in the profoundly deaf using a pitch extractor and storage oscilloscope display. IEEE *Transactions on Audio and Electroacoustics,* 1973, *21,* 274–278.

Branigan, G. H. Some early constraints on word combinations. *Dissertation Abstracts International.* 1977, *38,* 1988.

Breckenridge, J. Declination as a phonological process. *Bell Laboratories Technological Memo,* Murray Hill, N.J., 1977.

Breckenridge, J., and Liberman, M. Y. The declination effect in perception. Unpublished Manuscript, 1977. (Available from Bell Laboratories, Murray Hill, N.J.)

Bresnan, J. W. Sentence stress and syntactic transformations. *Language,* 1971, *47,* 257–281.

Brown, B. L., Strong, W. J., and Rencher, A. C. Fifty-four voices from two: The effects of simultaneous manipulations of rate, mean fundamental frequency, and variance of fundamental frequency on ratings of personality from speech. *Journal of the Acoustical Society of America,* 1974, *55,* 313–318.

Brown, R. and Kulik, J. Flashbulb memories. *Cognition,* 1977, *5,* 73–99.

Bruce, G. Swedish accents in sentence perspective. *Working Papers, Phonetics Laboratory,* Lund University, 1975, *12,* 61–70.

Caldwell, J. Flexible, high-performance speech synthesizer using custom NMOS circuitry. *Journal of the Acoustical Society of America,* 1978, *64,* S72 (Abstract).

Carlson, R., Erikson, Y., Granstrom, B., Lindblom, B., and Rapp, K. Neutral and emphatic stress patterns in Swedish. *Papers from the Institute of Linguistics,* University of Stockholm, 1974, *23,* 27–41.

Carlson, R., and Granstrom, B. Word accent, emphatic stress, and syntax in a synthesis by rule scheme for Swedish. *Stockholm Transmission Laboratory Quarterly Progress Status Report,* 1973, *2–3,* 31–36.

Carlson, R., Granstrom, B., Lindblom, B., and Rapp. K. Some timing and fundamental frequency characteristics of Swedish sentences: Data, rules, and a perceptual evaluation. *Stockholm Transmission Laboratory Quarterly Progress and Status Report,* 1972, *4,* 11–19.

Chang, C. T. Tones and intonation in the Cheng-tu dialect. *Phonetica,* 1958, *2,* 59–85.

Cheng, C. C. English stresses and Chinese tones in Chinese sentences. *Phonetica,* 1968, *18,* 77–88.

Chevrie-Muller, C., Dodart, F., Seguier-Dermier, N., and Salmon, D. Etude des paramètres acoustiques de la parole au cours de la schizophrénie de l'adolescent. *Folia Phoniatrica,* 1971, *23,* 401–428.

Chistovich, L. A. Variations of the fundamental voice pitch as a discriminatory cue for consonants. *Soviet Physics-Acoustics,* 1969, *14,* 372–378.

Chomsky, N. *Syntactic Structures.* The Hague: Mouton and Co., 1957.

Chomsky, N. *Aspects of the Theory of Syntax.* Cambridge, Mass.: M.I.T. Press, 1965.

Chomsky, N. Remarks on nominalization. In R. Jacobs and P. S. Rosenbaum (Eds.), *Readings in English Transformational Grammar.* Waltham, Mass.: Ginn, 1970.

Choppy, C. Prosodic features connected to punctuations: Some cases with commas. *Journal of the Acoustical Society of America,* 1978, *63,* S85–S86 (Abstract).

Clements, G. N. Tone and syntax in Ewe. In D. J. Napoli (Ed.), *Tone, Stress, and Intonation.* Washington, D.C.: Georgetown University Press, 1977.

Cohen, A. Comparison of Dutch and English intonation contours in spoken news bulletins. *Institute for Perception Research,* 1970.

Cohen, A., and t'Hart, J. On the anatomy of intonation. *Lingua,* 1967, *19,* 177–192.

Coker, C. H., Umeda, N., and Browman, C. P. Automatic synthesis from ordinary English Text. IEEE *Transactions on Audio and Electroacoustics,* 1973, *UA-21,* 293–297.

Cole, R. A. Listening for mispronunciations: A measure of what we hear during speech. *Perception and Psychophysics,* 1973, *14,* 153–156.

Coleman, R. O. A comparison of the contributions of two voice quality characteristics to the perception of maleness and femaleness in the voice. *Journal of Speech and Hearing Research,* 1976, *19,* 168–180.

Collier, R. Physiological correlates of intonation patterns. *Journal of the Acoustical Society of America,* 1975, *58,* 249–255.

Collier, R., and t'Hart, J. A grammar of pitch movements in Dutch intonation. *IPO Annual Progress Report,* 1971, *6,* 17–21.

Collier, R., and t'Hart, J. Perceptual experiments on Dutch intonation. *Proceedings of the VIIth International Congress of Phonetic Sciences, Montreal, 1971.* The Hague: Mouton, 1972. Pp. 880–884.

Colton, R. H., and Steinschneider, A. Acoustic characteristics of first week infant cries and their relationship to sudden infant death syndrome. *Journal of the Acoustical Society of America,* 1979, *66* (Supplement 1), S65 (Abstract).

Contini, M., and Boë, L.-J. Contribution à l'étude quantitative de l'evolution de la fréquence laryngienne dans la phrase enonciative en français. *Bulletin de l'institut de Phonetique de Grenoble,* 1973, *2,* 77–92.

Cooper, W. E., Danly, M., and Hamby, S. Fundamental frequency (F_0) attributes in the speech of Wernicke's aphasics. In J. J. Wolf and D. H. Klatt (Eds.), *Speech Communication Papers Presented at the 97th Meeting of the Acoustical Society of America.* New York: Acoustical Society of America, 1979.

Cooper, W. E., Egido, C., and Paccia, J. M. Grammatical control of a phonological rule: Palatalization. *Journal of Experimental Psychology: Human Perception and Performance,* 1978, *4,* 264–272.

Cooper, W. E., Lapointe, S. G., and Paccia, J. M. Syntactic blocking of phonological rules in speech production. *Journal of the Acoustical Society of America,* 1977, *61,* 1314–1320.

Cooper, W. E., Paccia, J. M., and Lapointe, S. Hierarchical coding in speech timing. *Cognitive Psychology,* 1978, *10,* 154–177.

Cooper, W. E., and Paccia-Cooper, J. *Syntax and Speech.* Cambridge, Mass.: Harvard University Press, 1980.

Cooper, W. E., and Sorensen, J. M. Fundamental frequency contours at syntactic boundaries. *Journal of the Acoustical Society of America,* 1977, *62,* 682–692.

Cooper W. E., and Zurif, E. B. Aphasia: Information-processing in language production and reception. To appear in B. Butterworth (Ed.), *Language Production, Volume II.* New York: Academic Press, 1981.

Cornut, G., Riou-Bourret, V., and Louis, M. H. Contribution à l'étude de la roix parlée et chantée de l'enfant normal de 5 à 9 ans. *Folia Phoniatrica,* 1971, *23,* 381–389.

Cruttenden, A. An experiment involving comprehension of intonation in children from 7 to 10. *Journal of Child Language,* 1974, *1,* 221–231.

Crystal, D. *Prosodic Systems and Intonation in English.* Cambridge: Cambridge University Press, 1969.

Curry, E. T. An objective study of the pitch characteristics of the adolescent male voice. *Speech Monographs,* 1940, *7,* 48.

Curry, E. T., Snidecor, J. C., and Isshiki, N. Fundamental frequency characteristics of Japanese Asai speakers. *The Laryngoscope,* 1973, *83,* 1759–1763.

Damste, P. H. Voice change in adult women caused by virilizing agents. *Journal of Speech and Hearing Disorders,* 1967, *32,* 126–132.

Danly, M., deVilliers, J. G., and Cooper, W. E. Control of speech prosody in Broca's aphasia. In J. J. Wolf and D. H. Klatt (Eds.), *Speech Communication Papers Presented at the 97th Meeting of the Acoustical Society of America,* 1979. Pp. 259–263.

Darwin, C. J., and Bethell-Fox, C. E. Pitch continuity and speech source attribution. *Journal of Experimental Psychology: Human Perception and Performance,* 1977, *3,* 665–672.

Delattre, P. Comparing the prosodic features of English, German, Spanish, and French. *International Review of Applied Linguistics (Heidelberg),* 1963, *1,* 193–210.

Delattre, P. The ten fundamental intonations in French. *French Review,* 1966, *15,* 1–14.

Delattre, P., Olsen, C., and Poenack, E. A comparative study of declarative intonation in American English and Spanish. *Hispania,* 1962, *45,* 233–241.

Delattre, P., Poenack, E., and Olsen, C. Some characteristics of German intonation for the expression of continuation and finality. *Phonetica,* 1965, *13,* 134–161.

Delgutte, B. Technique for the perceptual investigation of F_0 contours with application to French. *Journal of the Acoustical Society of America,* 1978, *64,* 1319–1332.

De Rooij, J. J. Speech punctuation: An acoustic and perceptual study of some aspects of speech prosody in Dutch. Unpublished Ph.D. Thesis, 1979.

Doherty, E. T. An evaluation of selected acoustic parameters for use in speaker identification. *Journal of Phonetics,* 1976, *4,* 321–326.

Donahue, M. L. Prosody and conversational turn-taking. Paper presented at the First Annual Boston University Conference on Language Development, Boston, Mass., 1976.

Downing, B. Parenthetical rules and obligatory phrasing. *Papers in Linguistics,* 1973, *6,* 108–128.

Duffy, R. J. Fundamental frequency characteristics of adolescent females. *Language and Speech,* 1970, *13,* 14–24.

Eckman, P., Friesen, W. V., and Scherer, K. R. Body movement and voice pitch in deceptive interaction. *Semiotica,* 1976, *16,* 23–27.

Efremova, I. B., Fintoft, K., and Ormestad, H. An experimental study of tonic accents in East Norwegian. *Norsk Tidsskrift for Sprogvidenskap,* 1965, *20,* 5–17.

Emonds, J. E. Parenthetical clauses. In C. Cornum, T. Smith-Stark, and A. Weiser (Eds.), *You Take the High Node and I'll Take the Low Node.* Chicago: Chicago Linguistic Society, 1973.

Emonds, J. E. *A Transformational Approach for Syntax: Root, Structure-Preserving, and Local Transformations.* New York: Academic Press, 1976.

Engen, T., and Ross, B. M. Long-term memory of odors with and without verbal descriptions. *Journal of Experimental Psychology,* 1973, *100,* 221–227.

Erikson, Y. Coordination of fundamental frequency and articulation in Swedish. *Working Papers, Phonetics Laboratory,* Lund University, 1973a, *9,* 59.

Erikson, Y. Preliminary evidence of syllable locked temporal control of F_0. *Stockholm Transmission Laboratory Quarterly Progress and Status Report,* 1973b, *2/3,* 23–30.

Erikson, Y., and Alstermark, M. Fundamental frequency correlates of the grave word accent in Swedish: The effect of vowel duration. *Stockholm Transmission Laboratory Quarterly Progress Status Report,* 1972, *2–3,* 53–60.

Es'kov, E. K. Vocal apparatus of the bees *Apis millifera* L. *Biofizika,* 1969, *14,* 158–166.

Ewan, W. G., and Krones, R. A study of larynx height in speech using the thyroumbrometer. *Journal of the Acoustical Society of America,* 1973, *53,* 345 (Abstract).

Ewan, W. G., and Krones, R. Measuring larynx movement using the thyro-umbrometer. *Journal of Phonetics,* 1974, *2,* 327–335.

Fant, G., Ishizaka, K., Lindqvist, J., and Sundberg, J. Spectral effects of subglottal cavity system. *Journal of the Acoustical Society of America,* 1972, *52,* 145.

Fernald, A. Rhythm and intonation in mothers' speech to newborns. Unpublished Paper. University of Oregon, 1978.

Ferreri, G. Senescence and the larynx. *Italian General Review of Oto-rhino-laryngology,* 1959, *1,* 640–709.

Filter, M. D., and Hyman, M. Relationship of acoustic parameters and perceptual ratings of esophageal speech. *Perceptual and Motor Skills,* 1975, *40,* 63–68.

Fintoft, K. Acoustical analysis and perception of tonemes in some Norwegian dialects. Universitetsforlaget, Oslo, 1970.

Flanagan, J. L., Coker, C. H., Rabiner, L. R., Schaefer, R. W., and Umeda, N. Synthetic voices for computers. *IEEE Spectrum,* 1970, *7.10,* 22–45.

Fonagy, I. Synthese de l'ironie. *Phonetica,* 1971, *23,* 42–51.

Fromkin, V. A. *Tone: A Linguistic Survey.* New York: Academic Press, 1978.

Fry, B. D. Duration and intensity as physical correlates of linguistic stress. *Journal of the Acoustical Society of America,* 1955, *27,* 765–768.

Fujimura, O. Remarks on stop consonants—Synthesis experiments and acoustic cues. In L. L. Hammerich, R. Jakobson, and E. Zwirner (Eds.), *Form and Substance: Phonetic and Linguistic Papers Presented to Eli Fisher-Jorgensen.* Copenhagen: Akademisk Forlag, 1971.

Fujimura, O. Stereo-fiberscope. In M. Sawashima and F. S. Cooper (Eds.), *Dynamic Aspects of Speech Production.* Tokyo: University of Tokyo Press, 1977. Pp. 133–138.

Fujisaki, H., Hirose, H., and Sugito, M. Analysis, synthesis, and perception of word accent types in Japanese. *Annual Bulletin, Research Institute of Logopedics and Phoniatrics,* University of Tokyo, 1976, *10,* 173–176.

Fujisaki, H., and Nagashima, S. A model for the synthesis of pitch contours of connected speech. *Annual Report, Engineering Research Institute, Faculty of Engineering,* University of Tokyo, 1967, *28,* 53–60.

Fujisaki, H., and Omura, T. Characteristics of duration of pauses and speech segments in connected speech. *Annual Report, Engineering Research Institute, Faculty of Engineering,* University of Tokyo, 1971, *30,* 69–74.

Fujisaki, H., and Sudo, H. Models for the word and sentence pitch contours of Japanese. *Annual Report, Engineering Research Institute, Faculty of Engineering,* University of Tokyo, 1970, *29,* 215–221.

Fujisaki, H., and Sudo, H. Synthesis by rule of prosodic features of connected Japanese. *International Congress on Acoustics,* 1971, 7.

Fujisaki, H., and Sugito, M. Acoustic and perceptual analysis of two-mora word accent types in Osaka dialect. *Annual Bulletin, Research Institute of Logopedics and Phoniatrics,* University of Tokyo, 1976, *10,* 157–171.

Gandour, J. Consonant types and tone in Siamese. *Journal of Phonetics,* 1974a, *2,* 337–350.

Gandour, J. On the representation of tone in Siamese. *Working Papers in Phonetics,* U.C.L.A., 1974b, 118–146.

Gårding, E. The Scandinavian word accents. *Working Papers, Phonetics Laboratory,* Lund University, 1973, 8.

Gårding, E. Swedish prosody by rule. Unpublished Paper, 1977a.

Gårding, E. The importance of turning-points for the pitch patterns of Swedish accents. In L. M. Hyman (Ed.), *Studies in Stress and Accent.* Los Angeles: University of Southern California, 1977b. Pp. 27–35.

Gårding, E. Sentence intonation in Swedish. *Phonetica,* 1979, *36,* 207–215.

Gårding, E., Fujimura, O., and Hirose, H. Laryngeal control of Swedish word tone—A preliminary report on an EMG study. *Annual Bulletin, Research Institute of Logopedics and Phoniatrics,* University of Tokyo, 1970, *4,* 45–54.

Gårding, E., and Lindblad, P. Constancy and variation in Swedish word accent patterns. *Working Papers, Phonetics Laboratory,* Lund University, 1973, *7,* 36–110.

Garnica, O. K. Some prosodic and paralinguistic features of speech to young children.

In C. E. Snow and C. A. Ferguson (Eds.), *Talking to Children: Language Input and Acquisition,* N.Y.: Cambridge University Press, 1977. Pp. 63–88.

Ghiselli, W. B., and LaRivière, C. Characteristics of ultrasonic vocalizations emitted by rats during shock-elicited aggression. *Animal Learning and Behavior,* 1977, *5,* 199–202.

Gilbert, H. Speech characteristics of miners with Black Lung Disease (Pneumacomiosis). *Journal of Communication Disorders,* 1975, *8,* 129–140.

Gilbert, H. R., and Weismer, G. G. The effect of smoking on the speaking fundamental frequency of adult women. *Journal of Psycholinguistic Research,* 1974, *3,* 225–231.

Gleason, J. Berko, and Weintraub, S. Input language and the acquisition of communicative competence. In K. Nelson (Ed.), *Children's Language. Volume I.* N.Y.: Gardner Press, 1978.

Gold, B., and Rabiner, L. R. Parallel processing techniques for estimating pitch periods of speech in the time-domain. *Journal of the Acoustical Society of America,* 1969, *46,* 442–448.

Goldsmith, J. An overview of autosegmental phonology. *Linguistic Analysis,* 1976, *2,* 23–68.

Goldstein, O. Comparison of terminal-analog synthesis of male and female voices. Unpublished S.B. Thesis, M.I.T., Cambridge, Mass., 1972.

Golub, H. L. A physioacoustic model of the infant cry and its use for medical diagnosis and prognosis. *Journal of the Acoustical Society of America,* 1979, *65,* S25 (Abstract).

Gonzalez, A. Acoustic correlates of accent, rhythm, and intonation in Tagalog. *Phonetica,* 1970, *22,* 11–44.

Goodglass, H., and Kaplan, E. *The Assessment of Aphasia and Related Disorders.* Philadelphia: Lea and Febiger, 1972.

Grosjean, F., and Collins, M. Breathing, pausing, and reading. *Phonetica,* 1979, *36,* 98–114.

Grosjean, F., Grosjean, L., and Lane, H. The patterns of silence: Performance structures in sentence production. *Cognitive Psychology,* 1979, *11,* 58–81.

Haavel, R. On the significance of some parameters of a pitch contour. In G. Fant and M. A. A. Tatham (Eds.), *Auditory Analysis and Perception of Speech.* New York: Academic Press, 1975. Pp. 511–520.

Hadding-Koch, K. Acoustico-phonetic studies in the intonation of Southern Swedish. Gleerup, Lund, 1961.

Hadding-Koch, K., and Studdert-Kennedy, M. A study of semantic and psycho-physical test responses to controlled variations in fundamental frequency. *Studia Linguistica,* 1963, *XVII,* 65–76.

Hadding-Koch, K., and Studdert-Kennedy, M. An experimental study of some intonation contours. *Phonetica,* 1964a, *11,* 175–184.

Hadding-Koch, K., and Studdert-Kennedy, M. Intonation contours evaluated by American and Swedish test subjects. Proceedings of the 5th International Congress of Phonetic Sciences, Munster (Basel-New York: S. Karger), 1964b, 326–331.

Hadding-Koch, K., and Studdert-Kennedy, M. Are you asking me, telling me, or talking to yourself? *Journal of Phonetics,* 1974, *2,* 7–14.

Haggard, M., Ambler, S., and Callow, M. Pitch as a voicing cue. *Journal of the Acoustical Society of America,* 1970, *47,* 613–617.

Haggard, M. P., and Parkinson, A. M. Stimulus and task factors as determinants of ear advantages. *Quarterly Journal of Experimental Psychology,* 1971, *23,* 158–177.

Halle, M., and Stevens, K. N. A note on laryngeal features. *Quarterly Progress Report of the M.I.T. Research Laboratory of Electronics,* 1971, *101,* 198–213.

Halliday, M. A. K. The tones of English. *Archivum Linguisticum,* 1963, *15,* 1–28.

Halliday, M. A. K. Intonation in English grammar. *Transactions of the Philological Society,* 1964, 143–169.

Halliday, M. A. K. Intonation systems in English. In A. MacIntosh and M. A. K. Halliday (Eds.), *Patterns of Language: Papers in General, Descriptive and Applied Linguistics.* London: Longmans, 1966. Pp. 111–133.

Halliday, M. A. K. *Intonation and Grammar in British English.* The Hague: Mouton, 1967.

Hankamer, J. Constraints on deletion in syntax. Unpublished Ph.D. Thesis. Yale University, New Haven, Conn., 1971.

Hankamer, J., and Sag, I. Deep and surface anaphora. *Linguistic Inquiry,* 1976, *7,* 391–426.

Hardcastle, W. J. Some observations on the *tense-lax* distinction in initial stops in Korean. *Journal of Phonetics,* 1973, *1,* 263–272.

Hartvigson, H. A specific case of terminal juncture and syntactic cohesion, *Phonetica,* 1965, *13,* 227–251.

Hauser, K. O. The use of acoustical analysis for identification of client stress within the counseling session. *Dissertation Abstracts International,* 1976, *36,* 5149–5150.

Hecker, M. H., Stevens, K. N., von Bismarck, G., and Williams, C. E. Manifestations of task-induced stress in the acoustic speech signal. *Journal of the Acoustical Society of America,* 1968, *44,* 993–1001.

Henke, W. L. Signals from external accelerometers during phonation: Attributes and their internal physical correlates. *Quarterly Progress Report, M.I.T. Research Laboratory of Electronics,* 1974, *114,* 224–231.

Hiki, S. On the control rules of voice pitch for sentence speech synthesis. *Journal of the Acoustical Society of Japan,* 1966, *22,* 364–367.

Hiki, S., Sugawara, K., and Oizumi, J. On the rapid fluctuation of voice pitch. *Journal of the Acoustical Society of Japan,* 1966, *22,* 290–291.

Hinton, G. A computational model of visual imagery. To appear in A. Baddeley and J. Long (Eds.) *Attention and Performance IX.* Hillsdale, N.J.: Lawrence Erlbaum Associates, 1981.

Hirano, M. Morphological structures of the vocal cord as a vibrator and its variations. *Folia Phoniatrica,* 1974, *26,* 89–94.

Hirose, H. On Japanese pitch accent: An acoustic analysis. Monthly Internal Memorandum, Phonology Laboratory, Berkeley, Calif., May 1971, 16–40.

Hirose, H., and Gay, T. The activity of the intrinsic laryngeal muscles in voicing control: An electromyographic study. *Phonetica,* 1972, *25,* 140–164.

Hirose, H., and Ushijima, T. Laryngeal control for voicing distinction in Japanese consonant production. *Phonetica,* 1978, *35,* 1–10.

Hixon, T. J., Klatt, D. H., and Mead, J. Influence of forced transglottal pressure changes on vocal fundamental frequency. *Journal of the Acoustical Society of America,* 1971, *49,* 105.

Hollien, H., and Hollien, P. A. A cross-cultural study of adolescent voice change in European males. *International Congress of Phonetic Sciences,* 1971, *7,* 119.

Hollien, H., Johnson, C. C., and Doherty, E. T. Speaker identification: New vectors for SAUSI. *Journal of the Acoustical Society of America,* 1978, *64,* S182 (Abstract).

Hollien, H., and Shipp, T. Speaking fundamental frequency and chronologic age in males. *Journal of Speech and Hearing Research,* 1972, *15,* 155–159.

Hollien, P. A., Brown, W. S., and Hollien, H. Speaking fundamental frequency (SFF) characteristics of children. *Journal of the Acoustical Society of America,* 1978, *64,* S90 (Abstract).

Hood, R. B., and Dixon, R. F. Physical characteristics of speech rhythm of deaf and normal-hearing speakers. *Journal of Communication Disorders,* 1969, *2,* 20–28.

Hoops, R. H., and Noll, D. J. Relationships of selected acoustic variables to judgments of esophageal speech. *Journal of Communication Disorders,* 1969, *2,* 1–13.

Horii, Y., and Ryan, W. J. Fundamental frequency characteristics and perceived age of adult male speakers. *Journal of the Acoustical Society of America,* 1975, *57,* S69 (Abstract).

Houde, R. A. Instantaneous visual feedback in speech training for the deaf. Paper presented at Convention of American Speech and Hearing Association, October, 1973.

House, A. S., and Fairbanks, G. The influence of consonant environment upon the secondary acoustical characteristics of vowels. *Journal of the Acoustical Society of America,* 1953, *25,* 105–113.

Howie, J. M. On the domain of tone in Mandarin: Some acoustical evidence. *Phonetica,* 1974, *30,* 129–148.

Hudgins, C. V., and Stetson, R. H. Voicing of consonants by depression of the larynx. *Archives neerlandaises de phonetique experimentale,* 1935, *11,* 1–28.

Huggins, A. W. F. A facility for studying perception of timing intonational speech. *Quarterly Progress Report, M.I.T. Research Laboratory of Electronics,* 1969, *95,* 81 83.

Huggins, A. W. F. An effect of syntax on syllable timing. *Quarterly Progress Report, M.I.T. Research Laboratory of Electronics,* 1974, *114,* 179–185.

Huttar, G. L. Relations between prosodic variables and emotions in normal American English utterances. *Journal of Speech and Hearing Research,* 1968, *11,* 481–487.

Hydrick, B. M., and Doddington, G. R. Performance evaluation of speaker verification in entry control. *Journal of the Acoustical Society of America,* 1978, *64,* S182 (Abstract).

Ichikawa, A., Nakajima, A., and Nakata, K. Speaker verification from actual telephone voice. *Journal of the Acoustical Society of America,* 1978, *64,* S182 (Abstract).

Isačenko, A., and Schädlich, H.-J. *A Model of Standard German Intonation,* translated by J. Philby. The Hague: Mouton, 1970.

Ishizaka, K., and Flanagan, J. L. Synthesis of voiced sounds from a two-mass model of the vocal cords. *The Bell System Technical Journal,* 1972, *51,* 1233–1268.

Ishizaka, K., and Matsudaira, M. What makes the vocal cords vibrate. In *6th International Congress of Acoustics,* Tokyo (Aug., 1968), 764–769.

Ivić, P., and Lehiste, I. Prilozi ispitivanju fonetskei fonoloske prirode akcenata u savremenom srpskohrvatskom knjizevnom jeziku. *Zbornik za filologiju i lingvistiku* (Novi Sad), 1963, *6,* 33–73.

Ivić, P., and Lehiste, I. Prilozi ispitivanju fonetskei fonoloske prirode akcenata u savremenom srpskohrvatskom jeziku, II. *Zbornik za filologiju i lingvistiku,* 1965, *8,* 75–117.

Ivić, P., and Lehiste, I. Prilozi ispitivanju fonetskei fonoloske prirode akcenata u savremenom srpskohrvatskom jeziku, III. *Zbornik za filologiju i lingvistiku,* 1967, *10,* 55–93.

Ivić, P., and Lehiste, I. Prilozi ispitivanju fonetskei fonoloske prirode akcenata u savremenom srpskohrvatskom jeziku, IV. *Zbornik za filologiju i lingvistiku,* 1969, *12,* 115–165.

Jackendoff, R. S. Gapping and related rules. *Linguistic inquiry,* 1971, *2,* 21–35.

Jackendoff, R. S. *$\overline{X}$ Syntax: A Study of Phrase Structure.* Cambridge, Mass.: M.I.T. Press, 1977.

Janota, P., and Liljencrants, J. The effect of fundamental frequency changes on the perception of stress by Czech listeners. *Stockholm Transmission Laboratory Quarterly Progress Status Report,* 1969, *4,* 32–38.

Jassem, W. Normalization of F_0 curves. In G. Fant and M. A. A. Tatham (Eds.), *Auditory Analysis and Perception of Speech.* New York: Academic Press, 1975. Pp. 523–530.

Jassem, W., Morton, J., and Steffen-Batog, M. The perception of stress in syntactic speech-like stimuli by Polish listeners. *Speech Analysis and Synthesis, (Warsaw),* 1968, *1,* 289–308.

Jeel, V. An investigation of the fundamental frequency of vowels after various Danish consonants, in particular stop consonants. *Annual Report of the Institute of Phonetics,* University of Copenhagen, 1975, *9,* 191–211.

Kagaya, R. Laryngeal gestures in Korean stop consonants. *Annual Bulletin, Research Institute of Logopedics and Phoniatrics,* University of Tokyo, 1971, *5,* 15–23.

Kagaya, R. A fiberscope and acoustic study of the Korean stops, affricates, and fricatives. *Journal of Phonetics,* 1974, *2,* 161–180.

Kagaya, R., and Hirose, H. Fiberoptic, electromyographic and acoustic analysis of Hindi stop consonants. *Annual Bulletin, Research Institute of Logopedics and Phoniatrics,* University of Tokyo, 1975, *9,* 27–46.

Kakita, Y., and Hiki, S. Thyrometer—A method for observing laryngeal control in speech. *Journal of the Acoustical Society of Japan,* 1972, *28,* 252–253.

Kakita, Y., and Hiki, S. A study of laryngeal control for voice pitch based on anatomical model. Eighth International Congress on Acoustics, July, 1974, London.

Kakita, Y., and Hiki, S. Investigation of laryngeal control in speech by use of a thyrometer. *Journal of the Acoustical Society of America,* 1976, *59,* 669–674.

Kersta, L. G. Voiceprint identification. *Nature,* 1962, *196,* 1253–1257.

Kim, C. W. On the autonomy of the tensity feature in stop classification (with special reference to Korean stops). *Word,* 1965, *21,* 339–359.

Klatt, D. H. Review of the ARPA Speech Understanding Project. *Journal of the Acoustical Society of America,* 1978, *62,* 1345–1366.

Klatt, D. H. Synthesis by rule of segmental durations in English sentences. In B. Lindblom and S. Öhman (Eds.), *Frontiers of Speech Communication Research.* N.Y.: Academic Press, 1979. Pp. 287–299.

Klatt, D. H., and Cooper, W. E. Perception of segment duration in sentence contexts. In A. Cohen and S. G. Nooteboom (Eds.), *Structure and Process in Speech Perception.* N.Y.: Springer-Verlag, 1975. Pp. 69–89.

Klatt, D. H., and Stefanski, R. A. How does a mynah bird imitate human speech? *Journal of the Acoustical Society of America,* 1974, *55,* 822–832.

Klein, W., Plomp, R., and Pols, L. C. W. Vowel spectra, vowel spaces, and vowel identification. *Journal of the Acoustical Society of America,* 1970, *48,* 999–1009.

Kosslyn, S. M. *Image and Mind.* Cambridge, Mass.: Harvard University Press, 1980.

Krovnova, O. F. Some notes on the melody component of intonation as a cue in text segmentation (summary). In G. Fant and M. A. A. Tatham (Eds.), *Auditory Analysis and Perception of Speech.* New York: Academic Press, 1975. P. 497.

Kuno, S. *The Structure of the Japanese Language.* Cambridge, Mass.: M.I.T. Press, 1973.

Kuzniarz, J. Basic frequencies for understanding Polish language in quietude. *Otolaryng. Pol.,* 1967, *21,* 269–276.

Kvavik, K. H. An analysis of sentence-initial and final intonational data in two Spanish dialects. *Journal of Phonetics,* 1974, *2,* 351–361.

Ladefoged, P. A phonetic study of West African languages: An auditory-instrumental survey. *West African Monograph Series,* 1964, *1,* 74 (Cambridge, England: Cambridge University Press, in association with the West African Language Survey).

Ladefoged, P. The features of the larynx. *Journal of Phonetics,* 1973, *1,* 73–83.

Ladefoged, P., and Broadbent, D. Information conveyed by vowels. *Journal of the Acoustical Society of America,* 1957, *29,* 98-104.

Lane, H., and Grosjean, F. Perception of reading rate by listeners and speakers. *Journal of Experimental Psychology,* 1973, *97,* 141–147.

Langacker, R. W. Movement rules in functional perspective. *Language,* 1974, *50,* 630–664.

Larreau, D., and Boe, L.-J. Etude de l'influence des variations de la frequence laryngienne sur l'intelligibilité et la qualité des consonnes sonores generées par vocodeur. *Bulletin de l'institut de Phonetique de Grenoble,* 1973, *2,* 103–126.

Laufer, M. Z., and Horii, Y. Fundamental frequency characteristics of infant nondistress vocalization during the first twenty-four weeks. *Journal of Child Language,* 1977, *4,* 171–184.

Lavelle, C. R. An experimental study of Yoruba tone. *Working Papers in Phonetics,* U.C.L.A., 1974, 160–170.

Lea, W. A. Intonational cues to the constituent structure and phonemics of spoken English. Unpublished Ph.D. Thesis, Purdue University, Lafayette, Ind., 1972.

Lea, W. A. Segmental and suprasegmental influences on fundamental frequency contours. In L. M. Hyman (Ed.), *Consonant Types and Tone.* Los Angeles: U.S.C. Press, 1973a. Pp. 15–70.

Lea, W. A. An approach to syntactic recognition without phonemics. *IEEE Transactions on Audio and Electroacoustics,* 1973b, *AU-21,* 249–258.

Lea, W. A. Sentences for controlled testing of acoustic phonetic components of speech understanding systems. *Sperry Univac Report No. PX 10952,* September 20, 1974.

Lea, W. A. Prosodic aids to speech recognition. In W. A. Lea (Ed.), *Trends in Speech Recognition.* Englewood Cliffs, N.J.: Prentice Hall, 1979, in press.

Lea, W. A., Medress, M., and Skinner, T. A prosodically-guided speech understanding strategy. *IEEE Transactions on Acoustics, Speech, and Signal Processing,* 1975, *ASSP-23,* 30–38.

Leben, W. R. The tones in English intonation. *Linguistic Analysis,* 1976, *2,* 69–107.

Lehiste, I. Some acoustic correlates of accent in Serbo-Croatian. *Phonetica,* 1961, *7,* 114–147.

Lehiste, I. *Suprasegmentals.* Cambridge, Mass.: M.I.T. Press, 1970.

Lehiste, I. Experiments with syntactic speech concerning quantity in Estonian. *Ohio State University Working Papers in Linguistics,* 1971, *9,* 199–217.

Lehiste, I. The phonetic structure of paragraphs. In A. Cohen and S. G. Nooteboom (Eds.), *Structure and Process in Speech Perception.* N.Y.: Springer-Verlag, 1975. Pp. 195–206.

Lehiste, I. Influence of fundamental frequency pattern on the perception of duration. *Journal of Phonetics,* 1976, *4,* 113–117.

Lehiste, I., and Ivic, P. Accent in Serbo-Croatian: An experimental study. *Michigan Slavic Materials 4* (Ann Arbor: University of Michigan), 1963.

Lehiste, I., and Ivic, P. Experiments with synthesized Serbo-Croatian tones. *Phonetica,* 1972, *26,* 1–15.

Lerman, J. W. Voice pitch of homosexuals. *Folia Phoniatrica,* 1969, *21,* 340–346.

Levelt, W. J. M., Zwanenburg, W., and Ouweneel, G. R. E. Ambiguous surface structure and phonetic form in French. *Foundations of Language,* 1970, *6,* 260–273.

Levin, H., and Lord, W. Speech pitch frequency as an emotional state indicator, *IEEE Transactions on Systems, Man, and Cybernetics,* 1975, *5,* 259–273.

Lhote, E. Apport de la glottospectographie a l'etude des tons. *International Congress of Phonetic Sciences,* 1971, *7,* 130.

Liberman, M. The intonational system of English. Unpublished Ph.D. Thesis, M.I.T., Cambridge, Mass., 1975.

Liberman, M., and Prince, A. On stress and linguistic rhythm. *Linguistic Inquiry,* 1977, *8,* 249–336.

Lieberman, P. On the acoustic basis of the perception of intonation by linguists. *Word,* 1965, *21,* 40–54.

Lieberman, P. *Intonation, Perception, and Language.* Cambridge, Mass.: M.I.T. Press, 1967.

Lieberman, P. Review of David Crystal's "Prosodic Systems and Intonation in English." Cambridge: Cambridge University Press, 1969. *Language,* 1976, *52,* 508–511.

Lightfoot, M. J. Accent and time in descriptive prosody. *Word,* 1970, *26,* 47–64.

Lilly, J. C. Distress call of the bottlenose dolphin: Stimuli and evoked behavioral responses. *Science,* 1963, *139,* 116–118.

Lilly, J. C. Vocal mimicry in *Tursiops.:* Ability to match number and duration of human vocal bursts. *Science,* 1965, *147,* 300–301.

Lind, J. Newborn infant cry. *Acta Paediat.,* Stockholm, Supplement 163, 1965.

Lindblom, B., and Rapp, K. Some temporal properties of spoken Swedish. *Papers from the Institute of Linguistics,* University of Stockholm, Publication 21, 1973.

Lindqvist, J. Laryngeal mechanisms in speech. *Stockholm Transmission Laboratory Quarterly Progress Status Report,* 1970, *2–3,* 26–32.

Lisker, L. On "explaining" vowel duration variation. *Glossa,* 1974, *2,* 233–246.

Lisker, L., and Abramson, A. S. A cross-language study of voicing in initial stops: Acoustical measurements. *Word,* 1964, *20,* 384–422.

Lisker, L., and Abramson, A. S. Some effects of context on voice onset-time in English stops. *Language and Speech,* 1967, *10,* 1–28.

Lisker, L., and Abramson, A. S. Distinctive features and laryngeal control. *Language,* 1971, *47,* 767–785.

Löfqvist, A. On intrinsic and extrinsic F_0 variations in Swedish tonal accents. *Working Papers, Phonetics Laboratory,* Lund University, 1973, *9.*

Löfqvist, A. Intrinsic and extrinsic F_0 variations in Swedish tonal accents. *Phonetica,* 1975, *31,* 228–247.

Lummis, R. C. Speaker verification: A step toward the "checkless" society. *Bell Laboratories Record,* 1972, *50,* 254–259.

Lyberg, B. Final lengthening—Partly a consequence of restrictions on the speed of fundamental frequency change? *Journal of Phonetics,* 1979, *7,* 187–196.

McAllister, R. Predicting physical aspects of English stress. *Stockholm Transmission Laboratory Quarterly Progress Status Report,* 1971, *1,* 20–29.

McCawley, J. *The Phonological Component of a Grammar of Japanese.* The Hague: Mouton, 1968.

McGarr, N. S., and Osberger, M. J. Pitch deviancy and intelligibility of deaf speech. *Journal of Communication Disorders,* 1978, *11,* 237–247.

Maeda, S. Electromyographic study on intonational attributes. *Quarterly Progress Report, M.I.T. Research Laboratory of Electronics,* 1975.

Maeda, S. A characterization of American English intonation. Unpublished Ph.D. Thesis, M.I.T., Cambridge, Mass., 1976.

Magdics, K. First findings in the comparative study of intonation of Hungarian dialects. *Phonetica,* 1964, *11,* 19–38, 101–115.

Magdics, K. Studies in the acoustic characteristics of Hungarian speech sounds. Indiana University, Bloomington. The Hague: Mouton, 1969.

Majewski, W., and Blasdell, R. Influence of fundamental frequency cues on the perception of some synthetic intonation contours. *Journal of the Acoustical Society of America,* 1969, *45,* 450–457.

Majewski, W., Hollien, H., and Zalewski, J. Speaking fundamental frequency of Polish adult males. *Phonetica,* 1972, *25,* 119–125.

Marslen-Wilson, W. D., and Teuber, H.-L. Memory for remote events in anterograde amnesia: Recognition of public figures from news photographs. *Neuropsychologia,* 1975, *13,* 353–364.

Massaro, D. W., and Cohen, M. M. Voice onset time and fundamental frequency as cues to the /zi/–/si/distinction. *Perception and Psychophysics,* 1977, *22,* 373–382.

Mattingly, I. G. Synthesis by rule of general American English. Unpublished Ph.D. Thesis, University of Connecticut, Storrs, 1968.

Menn, L., and Boyce, S. The course of fundamental frequency variations in parent's speech to children. Unpublished paper presented at the Linguistic Society Meeting, December, 1978.

Meyers, L. Tone patterns in Hausa: A re-analysis of Hausa downdrift. *Working Papers in Phonetics,* U.C.L.A., 1974, *27,* 47–62.

Meyers, L. Aspects of Hausa tone. *Working Papers in Phonetics,* U.C.L.A., 1976, *32,* 1–145.

Mikos, M. J. Intonation of questions in Polish. *Journal of Phonetics,* 1976, *4,* 247–253.

Mohr, B. Intrinsic variations on the speech signal. *Phonetica,* 1971, *23,* 65–93.

Monsen, R. B., and Engebretson, A. M. Study of variations in the male and female glottal wave. *Journal of the Acoustical Society of America,* 1977, *62,* 981–993.

Monsen, R., Engebretson, A., and Vemula, N. Indirect assessment of the contribution of subglottal air pressure and vocal-fold tension to changes of fundamental frequency in English. *Journal of the Acoustical Society of America,* 1978, *64,* 65–80.

Montague, J. C., Brown, W. S., and Hollien, H. Vocal fundamental frequency characteristics of institutionalized Down's syndrome children. *American Journal of Mental Deficiency,* 1974, *78,* 414–418.

Murray, T. Speaking fundamental frequency characteristics associated with voice pathologies. *Journal of Speech and Hearing Disorders,* 1978, *43,* 374–379.

Murray, T., Amundson, P., and Hollien, H. Acoustical characteristics of infant cries: Fundamental frequency. *Journal of Child Language,* 1977, *4,* 321–328.

Nash, R. *Turkish Intonation.* The Hague: Mouton, 1973.

Navarro, T. T. *Manual de Pronunciation Espanola.* 5th Edition. New York: Hafner Publishing Company, 1957.

Neeley, J. N., Edson, S. K., and Carlile, L. Speaking voice fundamental frequency of mentally retarded adults and normal adults. *American Journal of Mental Deficiency,* 1968, *72,* 944–947.

Nespor, M. A., and Allen, G. D. Segment and word durational correlates of syntactic boundaries in Italian. *Journal of the Acoustical Society of America,* 1977, *61,* S91 (Abstract).

Nickerson, R. S. Characteristics of the speech of deaf persons. *Volta Review,* 1975, *77,* 342–362.

Nooteboom, S. G., Brokx, J. P. L., and de Rooij, J. J. Contributions of prosody to speech perception. In W. J. M. Levelt and G. B. Flores d'Arcais (Eds.), *Studies in the Perception of Language,* 1978. Pp. 75–107.

Ohala, J., and Hirose, H. The function of the sternohyoid muscle in speech. *Annual Bulletin, Research Institute of Logopedics and Phoniatrics,* University of Tokyo, 1970, *4,* 41–44.

Öhman, S. E. G. Word and sentence intonation: A quantitative model. *Stockholm Transmission Laboratory Quarterly Progress Status Report,* 1967, *2–3,* 20–54.

Olive, J. P. Fundamental frequency rules for the synthesis of simple declarative English sentences. *Journal of the Acoustical Society of America,* 1975, *57,* 476–482.

Olsen, C. L. Voice register and intonation levels in two dialects of Spanish. Paper presented at the Modern Language Association, December, 1972.

O'Shaughnessy, D. User's manual for speech pitch extraction program. *M.I.T. Research Laboratory of Electronics, Natural Language Process Group, Memo 80* (unpublished), 1974.

O'Shaughnessy, D. Modelling fundamental frequency and its relationship to syntax, semantics, and phonetics. Unpublished Ph.D. Thesis, M.I.T., Cambridge, Mass., 1976.

O'Shaughnessy, D. Linguistic features in fundamental frequency patterns. *Journal of Phonetics,* 1979, *7,* 119–145.

Ostwald, P. The sounds of infancy. *Developmental Medicine and Child Neurology,* 1972, *14,* 350–361.

Paccia-Cooper, J., and Curcio, F. Language processing and forms of immediate echolalia in autistic children. *Journal of Speech and Hearing Research,* in press, 1980.

Painter, C. Archetypal breath-groups and the motor theory of speech perception: Evidence from a register tone language. *Anthropological Linguistics,* Oct., 1971, 349–360.

Painter, C. On the relationship between grammatical units, tonal units and physiological constraints on the respiratory system in GWA. *Zeitschrift tur Phonetik, Sprach Wissenschatt und Kommunikations forschung,* 1974, *27,* 302–319.

Painter, C. Pitch control and pharynx width in Twi: An electromyographic study. *Phonetica,* 1976, *33,* 334–352.

Parkhurst, B. G., and Levitt, H. The effect of selected prosodic errors on the intelligibility of deaf speech. *Journal of Communication Disorders,* 1978, *11,* 249–256.

Pernkopf, E. *Topographische Anatomie des Menschen,* Vol. 3. Wien-Innsbruck: Urban and Schwarzenberg, 1952.

Petersen, N. R. Intrinsic fundamental frequency of Danish vowels. *Journal of Phonetics,* 1978, *6,* 177–189.

Peterson, G. E., and Barney, H. L. Control methods used in a study of the vowels. *Journal of the Acoustical Society of America,* 1952, *24,* 175–184.

Pétursson, M. Intonationen i den enkle deklarative saetning i islandsk. In E. Gårding, G. Bruce, and R. Bannert (Eds.), *Nordic Prosody: Papers from a Symposium.* Department of Linguistics, Lund University, 1978. Pp. 33–42.

Pierrehumbert, J. Intonation synthesis based on metrical grids. In J. J. Wolf and D. H. Klatt (Eds.), *Speech Communication Papers Presented at the 97th Meeting of the Acoustical Society of America.* New York: Acoustical Society of America, 1979a.

Pierrehumbert, J. The perception of fundamental frequency declination. *Journal of the Acoustical Society of America,* 1979b, *66,* 363–369.

Purcell, E. T. Pitch peak location and the perception of Serbo-Croatian word tone. *Journal of Phonetics,* 1976, *4,* 265–270.

Purcell, E. T. The relative importance of initial/final pitch ratio and of pitch peak location for the perception of Serbo-Croatian tone. *Journal of the Acoustical Society of America,* 1978, *64,* S20 (Abstract).

Rabiner, L. R., Cheng, M. J., Rosenberg, A. E., and McGonegal, C. A. A comparative performance study of several pitch detection algorithms. *IEEE Transactions on Acoustics, Speech, and Signal Processing,* 1976, *ASSP-24,* 399–418.

Reddy, D. R. (Ed.), *Speech Understanding Systems.* New York: Academic Press, 1971.

Rehder, P. Beitrage zur Erforschung der Serbokroatischen Prosodie. *Slavistische Beitrage 31* (Munchen: Verlag Otto Sagner), 1968.

Rice, D. G., Abroms, G. M., and Saxman, J. H. Speech and physiological correlates of "flat" affect. *Archives of General Psychiatry,* 1969, *20,* 566–572.

Rigault, A. Role de la frequence, de l'intensité et de la durée vocaliques dans la perception de l'accent en français. *Proceedings of the 4th International Congress of Phonetic Sciences,* Helsinki (The Hague: Mouton and Co.), 1961, 735–748.

Robinson, D. F. Some acoustic correlates of tone in standard Lithuanian. *The Slavic and East European Journal,* 1968, *12,* 206–212.

Ross, J. R. A proposed rule of tree-pruning. In D. Riebel and S. Schame (Eds.), *Modern Studies in English: Readings in Transformational Grammar.* Englewood Cliffs, N.J.: Prentice-Hall, 1969.

Ross, J. R. Gapping and the order of constituents. In M. Bierwisch and K. E. Hiedolph (Eds.), *Progress in Linguistics.* The Hague: Mouton, 1970.

Rutter, M., and Schopler, E. (Eds.), *Autism: A Reappraisal of Concepts and Treatment.* New York: Plenum, 1978.

Sawashima, M., and Hirose, H. New laryngoscopic technique by use of fiber-optics. *Journal of the Acoustical Society of America,* 1968, *43,* 168–169.

Sawashima, M., Kakita, Y., and Hiki, S. Activity of the extrinsic laryngeal muscles in relation to Japanese word accent. *Annual Bulletin, Research Institute of Logopedics and Phoniatrics,* University of Tokyo, *7,* 1973.

Saxman, J. H., and Burk, K. W. Speaking fundamental frequency and rate characteristics of adult female schizophrenics. *Journal of Speech and Hearing Research,* 1968, *11,* 194–203.

Scherer, K. R. Voice quality analysis of American and German speakers. *Journal of Psycholinguistic Research,* 1974, *3,* 281–298.

Scherer, K. R., London, H., and Wolf, J. J. The voice of confidence: Paralinguistic cues and audience evaluation. *Journal of Research in Personality,* 1973, *7,* 31–44.

Schneider, H. The effect of local heating on the calls of the fire-bellied toad, *Bombina bombina. Behavioral Processes,* 1976, *1,* 135–144.

Sedlacek, K. Hearing and communication in birds: Species agapornus roseicolis. *Acta Oto-Laryngologica,* 1971, *71,* 194–196.

Serniclaes, W., and Wajskop, M. Vowel identification as a function of fundamental frequency and the duration of presentation. *Université Libre de Bruxelles, Rapport d'activites de l'Institut de Phonetique,* 1971, *4,* 54–70.

Shimaoka, T. A contrastive study on rhythm and intonation of English and Japanese with spectographic analysis. *Study of Sounds,* 1966, 12.

Shipp, T. Frequency, duration, and perceptual measures in relation to judgment of alaryngeal speech acceptability. *Journal of Speech and Hearing Research*, 1967, *10*, 417–427.

Slepokurova, N. A. Sensory systems; problems of theory and research methods in the perception of speech signals, No. 2: The influence of the frequency of the fundamental tone on the position of the phoneme boundary between the vowels /a/ and /o/. *Informatsionnye materialy, Akademiya Nauk. SSSR Ob'edinyonnyi nauchniyi Sovet Fiziologiya Cheloveka i Zhivotnykh*, 1971, *32*, 60–70.

Sonesson, B. The functional anatomy of the speech organs. In B. Malmberg (Ed.), *Manual of Phonetics*, Amsterdam: North-Holland Publishing Company, 1970.

Sorensen, J. M., and Cooper, W. E. Syntactic coding of fundamental frequency in speech. In R. A. Cole (Ed.), *The Perception and Production of Fluent Speech*. Hillsdale, N.J.: Lawrence Erlbaum Associates, 1980, Pp. 399–440.

Springer, S. P. Memory for linguistic and nonlinguistic dimensions of the same acoustic stimulus. *Journal of Experimental Psychology*, 1973, *101*, 159–163.

Sternberg, S., Wright, C. E., Knoll, R. L., and Monsell, S. Motor programs in rapid speech: Additional evidence. In R. A. Cole (Ed.), *Perception and Production of Fluent Speech*. Hillsdale, N.J.: Lawrence Erlbaum Associates, 1980. Pp. 507–534.

Stevens, K. N. Physics of laryngeal behavior and larynx modes. *Phonetica*, 1977, *34*, 264–279.

Stevens, K. N., Henke, W. L., and Sorensen, J. M. Unpublished data, M.I.T., Cambridge, Mass., 1976.

Stokes, W., and Branigan, G. On the definition of two word utterances: Or when does 1 + 1 = 2? Unpublished Paper, Boston University, 1978.

Stratton, W. D. Intonation feedback for the deaf through a tactile display. *Volta Review*, 1974, *76*, 26–35.

Streeter, L. A. Acoustic determinants of phrase boundary perception. *Journal of the Acoustical Society of America*, 1978, *64*, 1582–1592.

Streeter, L. A., Krauss, R. M., Apple, W., and MacDonald, N. H. Acoustic consequences and perceptual indicators of stress. *Journal of the Acoustical Society of America*, 1978, *64*, S115 (Abstract).

Studdert-Kennedy, M., and Hadding-Koch, K. Auditory and linguistic processes in the perception of intonation contours. *Language and Speech*, 1973, *16*, 293–313.

Studdert-Kennedy, M., and Shankweiler, D. Hemispheric specialization for speech perception. *Journal of the Acoustical Society of America*, 1970, *48*, 579–594.

Sudo, T. A study on a character of a speech trainer: I. Acoustic components and perception of intonation in speech. *Japanese Journal of Special Education*, 1973, *11*, 43–49.

Summerfield, A. Q., and Haggard, M. P. Speech rate effects in the perception of voicing. *Speech Synthesis and Perception: Research on Speech Synthesis and Speech Perception in the Psychology*, Cambridge University, 1972, *6*, 1–12.

Svetozarova, N. D. The inner structure of intonation contours in Russian. In G. Fant and M. A. A. Tathan (Eds.), *Auditory Analysis and Perception of Speech*. New York: Academic Press, 1975. Pp. 499–510.

Swanson, D. B. Vocal correlates of depression. *Dissertation Abstracts International*, 1977, *38*, 918–919.

Tatsumi, M., Kunisaki, O., and Fujisaki, H. Acoustic analysis and subjective evaluation of sung vowels. *Annual Bulletin, Research Institute of Logopedics and Phoniatrics*, University of Tokyo, 1976, *10*, 191–198.

Terrango, L. Pitch and duration characteristics of the oral reading of males on a masculinity–femininity dimension. *Journal of Speech and Hearing Research*, 1966, *9*, 590–595.

t'Hart, J., and Cohen, A. Intonation by rule: A perceptual quest. *Journal of Phonetics,* 1973, *1,* 309–327.

t'Hart, J., and Collier, R. A course in Dutch intonation. *IPO Annual Progress Report,* 1978, *13,* 31–35.

Thorsen, N. An acoustical investigation of Danish intonation. *Journal of Phonetics,* 1978a, *6,* 151–175.

Thorsen, N. Aspects of Danish intonation. In E. Gårding, G. Bruce, and R. Bannert (Eds.), *Nordic Prosody: Papers from a Symposium.* Department of Linguistics, Lund University, 1978b. Pp. 23–32.

Titze, I. R. On the mechanics of vocal-fold vibration. *Journal of the Acoustical Society of America,* 1976, *60,* 1366–1380.

Trager, G. L., and Smith, H. L. *Outline of English Structure, Studies in Linguistics, No. 3.* Norman, Okla.: Battenburg, 1951.

Trimboli, F. Changes in voice characteristics as a function of trait and state personality variables. *Dissertation Abstracts International,* 1973, *33,* 3965.

Utsuki, N., and Okamura, N. Relationship between emotional state and fundamental frequency of speech. Japanese Air Self Defense Force, Aeromedical Lab, Tokyo. *Report of Aeromedical Laboratory,* 1976, *16,* 179–188.

Vaissière, J. On French prosody. *Quarterly Progress Report, M.I.T. Research Laboratory of Electronics,* 1974, *114,* 212–224.

Vaissière, J. Nuclear stress rule in French. *Journal of the Acoustical Society of America,* 1975, *57,* S1 (Abstract).

Vaissière, J. Quelques experiences d'analyse perceptuelle en Français. *Actes des 8 emes Journées d'Étude sur la Parole, Groupement des Acousticiens de Langue Français, Aix-en-Provence,* 1977, 183–189.

van den Berg, J. W. Mechanisms of the larynx and the laryngeal vibrations. In B. Malmberg (Ed.), *Manual of Phonetics.* Amsterdam: North-Holland Publishing Company, 1970.

van Geel, R. C. Pitch in electrolarynx speech. *Progress Report, Institute of Phonetics,* University of Utrecht, The Netherlands, 1979, *4,* 27–35.

van Katwijk, A. The composition of pitch contours in Dutch utterances. *Institute for Perception Research,* 1970, *5.*

Wang, W. S-Y. Phonological features of tone. *International Journal of American Linguistics,* 1967, *33,* 93–105.

Wangler, H. *Zur Topologie des Hausa,* Sniften zur Phonetik, Sprachwissenschaft und kommunikationsforschung Nr. 6. Akademie-Verlag, Berlin, 1963.

Wasz-Hockert, O., Lind, J., Vuorenkoski, V., Partanen, T., and Valanne, E. *The Infant Cry: A Spectographic and Auditory Analysis.* London: Heinemann, 1968.

Watanabe, A., and Okamura, H. Speech trainer for correction of intonation and its effect to hard of hearing children. *Journal of the Acoustical Society of Japan,* 1976, *32,* 556–557.

Weaver, C. T. Characteristics of mothers' vocal pitch to younger and older children during mother–child interactions. *Dissertation Abstracts International,* 1976, *37,* 2534.

Weinberg, B., and Zlatin, M. Speaking fundamental frequency characteristics of five- and six-year-old children with mongolism. *Journal of Speech and Hearing Research,* 1970, *13,* 418–425.

Westin, K., Buddenhagen, R. G., and Obrecht, D. H. An experimental analysis of the relative importance of pitch, quantity, and intensity as cues to phonemic distinctions in Southern Swedish. *Language and Speech,* 1966, *9,* 114–126.

Willems, N. J. Discriminability of Dutch and English intonation contours. *Progress Report, Institute of Phonetics,* University of Utrecht, The Netherlands, 1978, *3,* 3–17.

Williams, C. E., and Stevens, K. N. Emotions and speech: Some acoustical correlates. *Journal of the Acoustical Society of America,* 1972, *52,* 1238–1250.

Wolf, C. G. The processing of fundamental frequency in a dichotic matching task. *Brain and Language,* 1977, *4,* 70–77.

Wolf, J. J. Efficient acoustic parameters for speaker recognition. *Journal of the Acoustical Society of America,* 1972, *51,* 2044–2056.

Wolff, P. The natural history of crying and other vocalizations in early infancy. In B. M. Foss (Ed.) *Determinants of Infant Behavior,* Vol. 4. London: Methuen, 1969. Pp. 81–111.

Yasuhiro, T., and Ozeki, K. An experiment on male to female voice conversion. *Journal of the Acoustical Society of Japan,* 1976, *32,* 362–368.

Yorifuji, K., Cooper, W. E., and Sorensen, J. M. Unpublished data, M.I.T., Cambridge, Mass., 1978.

Zee, E. Duration and intensity as correlates of F_0. *Journal of Phonetics,* 1978, *6,* 213–220.

Zeskind, P. S., and Lester, B. M. Acoustic features and auditory perceptions of the cries of newborns with prenatal and perinatal complications. *Child Development,* 1978, *49,* 580–589.

Zetterlund, S., Nordstrand, L., and Engstrand, O. An experiment on the perceptual evaluation of prosodic parameters for phrase structure decision in Swedish. In E. Gårding, G. Bruce, and R. Bannert (Eds.), *Nordic Prosody: Papers from a Symposium.* Department of Linguistics, Lund University, 1978. Pp. 15–22.

Zwanenburg, W. Rhythme et melodie de la phrase parlée en France et au Quebec. By M. Boudreault. Laval, Quebec: University Press Laval, 1968. *Lingua. Amsterdam,* 1970, *26,* 100–103.

Author Index

Subject Index